Lecture Notes in Computer Science   11991

More information about this series at http://www.springer.com/series/7412

Hassan Mohy-ud-Din · Saima Rathore (Eds.)

Radiomics and Radiogenomics in Neuro-oncology

First International Workshop, RNO-AI 2019
Held in Conjunction with MICCAI 2019
Shenzhen, China, October 13, 2019
Proceedings

 Springer

Editors
Hassan Mohy-ud-Din ⓘD
LUMS School of Science and Engineering
Lahore, Pakistan

Saima Rathore
University of Pennsylvania
Philadelphia, PA, USA

ISSN 0302-9743   ISSN 1611-3349   (electronic)
Lecture Notes in Computer Science
ISBN 978-3-030-40123-8   ISBN 978-3-030-40124-5   (eBook)
https://doi.org/10.1007/978-3-030-40124-5

LNCS Sublibrary: SL6 – Image Processing, Computer Vision, Pattern Recognition, and Graphics

This Springer imprint is published by the registered company Springer Nature Switzerland AG
The registered company address is: Gewerbestrasse 11, 6330 Cham, Switzerland

Preface

Due to the exponential growth of computational algorithms, AI methods are poised to improve the precision of diagnostic and therapeutic methods in medicine. The field of radiomics in neuro-oncology has been and will likely continue to be at the forefront of this revolution. A variety of AI methods applied to conventional and advanced neuro-oncology MRI data can do several tasks. The first essential step of radiomics generally involves lesion segmentation, which is generally preceded by image pre-processing steps including skull stripping, intensity normalization, and alignment of image volumes from different modalities. A variety of methods have been used for segmentation, ranging from manual labeling and/or annotation and semiautomated methods to more recent deep learning methods. The next step of radiomics with traditional machine learning involves the extraction of quantitative features, including basic shape, size, and intensity metrics, as well as more complex features derived from a variety of statistical approaches applied to the images, for example, histogram-based features, texture-based features, fitted biophysical models, spatial patterns, and deep learning features. A variety of different machine learning models can then be applied to the intermediate quantitative features in order to "mine" them for significant associations, allowing them to predict crucial information about a tumor, such as infiltrating tumor margins, molecular markers, and prognosis, which are relevant for therapeutic decision making. Alternatively, deep learning approaches to radiomics in neuro-oncology generally necessitate less domain-specific knowledge compared with the explicitly engineered features for traditional machine learning, allowing them to make predictions without explicit feature selection or reduction steps.

Radiogenomics has also advanced our understanding of cancer biology, allowing noninvasive sampling of the molecular environment with high spatial resolution and providing a systems-level understanding of underlying heterogeneous cellular and molecular processes. By providing in vivo markers of spatial and molecular heterogeneity, these AI-based radiomic and radiogenomic tools have the potential to stratify patients into more precise initial diagnostic and therapeutic pathways and enable better dynamic treatment monitoring in this era of personalized medicine. Although substantial challenges remain, radiologic practice is set to change considerably as AI technology is further developed and validated for clinical use.

The first edition of the Radiomics and Radiogenomics in Neuro-oncology using AI (RNO-AI 2019) workshop[1] was successfully held in conjunction with the 22nd International Conference on Medical Image Computing and Computer-Assisted Intervention (MICCAI 2019) in Shenzhen, China, on October 13, 2019. The aim of RNO-AI was to bring together the growing number of researchers in the field given the significant amount of effort in the development of tools that can automate the analysis and synthesis of neuro-oncologic imaging. Submissions were solicited via call for

---

[1] https://sites.google.com/view/rno-ai2019/.

papers by the MICCAI and workshop organizers, as well as by directly emailing more than 400 colleagues and experts in the area. Each submission underwent a double-blind review by at least three members of the Program Committee, consisting of researchers actively contributing in the area. Three invited papers were also solicited from leading experts in the field. RNO-AI 2019 featured three keynote talks and seven oral presentations. The duration of the workshop was approximately five hours.

We would like to extend our gratitude to the members of the Program Committee for their reviews; keynote speakers, Prof. Michel Bilello, Prof. Pallavi Tiwari, and Prof. Bjoern Menze, for illuminating talks; authors for their research contributions; and the MICCAI society for their overall support.

November 2019   · Hassan Mohy-ud-Din
                                                            Saima Rathore

Organization

Organizing Committee

Hassan Mohy-ud-Din   LUMS School of Science and Engineering, Pakistan
Saima Rathore   University of Pennsylvania, USA
Madhura Ingalhalikar   Symbiosis Institute of Technology, India

Program Committee

Ulas Bagci   University of Central Florida, USA
Bjoern Menze   Technische Universität München, Germany
Pallavi Tiwari   Case Western Reserve University, USA
Nicolas Honnorat   SRI International, USA
Ahmad Chaddad   McGill University, Canada
Zahra Riahi Samani   University of Pennsylvania, USA
Yusuf Osmanlioglu   University of Pennsylvania, USA

Contents

Current Status of the Use of Machine Learning and Magnetic Resonance Imaging in the Field of Neuro-Radiomics

Ashish Singh and Michel Bilello[✉]

Center for Biomedical Image Computing and Analytics (CBICA),
Department of Radiology, Perelman School of Medicine,
University of Pennsylvania, Philadelphia, PA, USA
michel.bilello@pennmedicine.upenn.edu

Abstract. Brain tumors exhibit heterogeneous profile with anomalous hemo-dynamics. In spite of significant advances at diagnostic and therapeutic fronts in the past couple of decades, the prognosis still remains poor. Magnetic resonance imaging (MRI), which provides information about the structural and functional aspects of the tumor in a noninvasive manner, has gained a lot of popularity for evaluating brain tumors. Several studies have been proposed in the recent past that focused on quantifying the characteristics of brain tumors as seen on MRI scans in terms of various descriptors, such as shape/morphology, texture, signal strength, and temporal dynamics, and then integrating these quantitative descriptors into various diagnostic and prognostic indices. This article first presents an overview of various MRI imaging sequences, such as contrast-enhanced, dynamic susceptibility contrast, diffusion tensor imaging, and con-ventional MRI, used in routine clinical settings. Later, it provides a detailed overview of the current status of the use of machine learning on MRI scans, with focus on clinical applications of these imaging sequences in brain tumors, including grading, assessment of the treatment response, prediction of progno-sis, and identification of molecular markers. The article also highlights current challenges and future research directions.

Keywords: Radiomics · Neuro-oncology · Magnetic resonance imaging · Machine learning · Brain tumors

1 Introduction

Adult brain tumors, starting from the least lethal pilocytic astrocytoma tumors and going beyond the most lethal glioblastoma (GBM) [1], show erratic proliferative potential and malignancy. Despite significant diagnostic and therapeutic advances made in the recent years, the survival outcome still remains poor, with median survival of ∼14 months for the most lethal brain tumors (GBM). Vascularization, with abnormal structure and functionality, is a typical hallmark of these tumors and that drives various biological behaviors, including progression, invasion, and confrontation with various therapeutic options [2]. Another challenge in the treatment of brain tumors is their inter- and intra-tumor heterogeneity, which poses several diagnostic and

© Springer Nature Switzerland AG 2020
H. Mohy-ud-Din and S. Rathore (Eds.): RNO-AI 2019, LNCS 11991, pp. 1–11, 2020.
https://doi.org/10.1007/978-3-030-40124-5_1

therapeutic challenges. An understanding of the structural and functional characteristics of these tumors might lead to improved brain tumor management.

Magnetic resonance imaging (MRI) is the most commonly used imaging modality for radiographic assessment of brain tumors [3]. In addition, it is also used to continuously monitor the dynamic changes happening in the tumor, such as determining extent of resection, assessing response of the treatment, determining progression level of the disease, during the treatment course, thereby facilitating the care of brain tumor patients [4–6]. MRI also provides an extended characterization of the tumors including anatomical location, morphological descriptors such as shape and size, and heterogeneity level of the tumor, which are currently being analyzed and quantified by the neuro-radiologist.

MRI, though, captures massive amount of information, but almost all the radiographic data are analyzed and reported in qualitative manner by the radiologists. Radiomics is an evolving field that converts clinical radiographic data, which is subjective, into a high dimensional comprehensive feature space using advanced medical image analysis algorithms designed for data-quantification [7, 8]. Radiomics has the combined advantages of being highly patient-specific and non-invasive. Moreover, radiomics helps quantifying the heterogeneity of the whole tumor and also facilitates the monitoring during the course of the disease in contrast to biopsy specimens which have several inherent limitations such as tissue-sampling error and lack of temporal assessment of the disease. There is a recent trend in combining radiologic features extracted from radiology data with the pathology features extracted from whole-slide images, which has proven to be effective compared to either of these alone [9].

This article presents a general overview of the role of radiomics in the field of neuro-oncology. It also sheds light on various MRI sequences currently acquired as part of routine clinical practice. Later, it provides a detailed account of the current status of the use of machine learning (ML) in the field of neuro-oncology, with a particular focus on glioblastoma, such as identification of molecular characteristics of the tumors, an assessment of the response of the treatment, and prediction of survival outcome. Current challenges the field is facing, and future research directions are provided at the end.

2   Neuro-Radiomics

Radiomics [8, 10] is all about the extraction of quantitative features from various sub-regions of the tumors with an aim to elucidate the underlying biological behaviors and improve the treatment of the patients. Machine learning algorithms are then applied on this data to find the hidden patterns inside the data and to validate these measures as quantitative biomarkers in order to summarize and understand the dynamics of a tumor throughout the management of the disease [11]. Quantitative features, extracted from the tumors, such as signal strength, morphological descriptors of shape and size, and texture measures provide a quantitative assessment of the phenotype and microenvironment of the tumors; this information is comprehensive and is quite distinct from the data provided by clinical record, test results, and genomic assays. These quantitative measures, along with the other clinical, genetic, and proteomic data, can be used to build diagnostic and predictive models [10].

Radiomics in brain tumors usually involves a series of image processing steps: (i) preprocessing of the images including stripping of the skull, normalization of the intensity values, and alignment of the images in a single orientation, (ii) segmentation of various sub-regions of the tumors using different methods such as manual [12], semi-automatic [13] or fully-automated methods [14], (iii) quantification of the MRI-based characteristics of tumors such as morphology (shape, size), intensity metrics, histogram profiling based features, patterns of spatial distribution, etc., and (iv) data mining using various available machine learning models to extract important information about a tumor, including infiltrating margins of the tumor, molecular characteristics of the tumor, and survival outcome which are important for therapeutic decision making.

3   Standard MRI Protocols

In Neuro-Radiomics, MRI plays an important role due to several reasons such as high soft-tissue contrast and availability of multi-parametric MRI (mpMRI) that non-invasively and non-destructively provides information for characterization of tumor and surrounding tissues. Furthermore, each sequence of MRI provides information on different tissue characteristics. Regional angiogenesis and integrity of blood-brain barrier in the tumor are highlighted in T1-weighted post-contrast (T1-Gd) sequence, information on *in-vivo* assessment of the extracellular fluid in brain parenchyma and tissue necrosis is provided by T2-weighted (T2) and T2-fluid attenuated inversion recovery (T2-FLAIR) sequences, the water diffusion process in the brain, affected in part by tumor cells architecture and density [15] is characterized by diffusion tensor imaging (DTI) while quantification of regional microvasculature and hemodynamics [16, 17] is made possible by dynamic susceptibility contrast-enhanced (DSC)-MRI techniques.

A lot of data has been made available via TCIA that supports the advances in the field of segmentation and radiomics. However, there has been very little work done in the field, and among them, a few have resulted in open source software such as Modelhub (www.modelhub.ai), 3D Slicer [18], PyRadiomics [19], CaPTk [20], etc.

4   Machine Learning in Neuro-Radiomics

Machine learning models are developed using quantitative information obtained from MRI sequences for diagnosis of tumors, patient prognosis [21], detection of relative tumor heterogeneity which later guides in clinical decision making [22], peritumoral heterogeneity/infiltration [23], and assessment of pseudo-progression [24]. We briefly describe and discuss these applications of radiomics in neuro-oncology in the following text.

4.1 Survival of Brain Tumors

Prediction of survival at the initial presentation of disease is important not only for patient management, but also for stratifying patients into clinical trials and for evaluating treatment effects of certain drugs/therapies. Many studies in the recent past have established robust and reproducible MRI-based imaging predictors of survival for brain tumors [3, 25, 26].

Among the studies using basic feature extraction methods, Macyszyn et al. [21] employed features of intensity, volume, and location of the tumor, explicitly extracted from traditional and advanced MRI sequences, and developed a support vector machine (SVM) model to predict survival groups (<6 months, >6 months, >6 and <18 months), achieving an accuracy of 80%. Rathore et al. used intensity features extracted from traditional MRI, which were normalized by the intensity values within the normal tissue on the contralateral side of the tumor. They applied the method on multi-institutional dataset and achieved quite robust survival prediction on multi-institutional data [27]. Several studies have utilized texture features for developing predictors of survival. Sanghani et al. [25] used a comprehensive set of tumor volumetric and shape features along with texture features derived from multi-channel MR images of glioblastoma patients for prediction of overall-survival. Patient age was also used as a feature in this study. Likewise, Yang et al. [26] developed a very robust predictor of survival, at a 12 months cut-off, by using different texture feature types. The usefulness of multi-scale texture MRI features for discriminating glioblastoma regions and predicting overall-survival has also been evaluated in the past [3].

The use of deep learning architectures has also received some attention for the prediction of survival of brain tumors. Chato et al. [28] used several machine learning and deep learning methods to predict overall-survival. Various types of features were extracted and trained using different machine learning techniques. The best classification accuracy was achieved by using deep feature extraction based on pre-trained AlexNet [29] and trained by Linear Discriminant. Similarly, Lao et al. [30] investigated if deep features extracted via transfer learning can generate radiomics signatures for prediction of overall-survival in patients with glioblastoma. Deep features were extracted from pre-trained convolutional neural network (CNN) via transfer learning. Their study demonstrated that transfer learning-based deep features were able to generate prognostic imaging signature for overall-survival prediction and patient stratification for glioblastoma.

A recent study [31] compared the predictive performance of machine learning based methods to human knowledge-based approaches. They developed a human-built optimized linear predictive model (OLPM) on the basis of the researchers' understanding of the predictive value of the variables showing individual prognosis value. They found that OLPM outperformed other machine learning models based on either the same parameter set or on the full set of 44 features considered (Table 1).

Table 1. Summary of the ML based radiomic signatures developed for prognostication of brain tumors

Study	Patients	ML algorithm	Accuracy
Macyszyn et al. [21]	29	SVM	79.17%
Chato et al. [28]	163	AlexNet	73.00%
Sanghani et al. [25]	163	SVM	98.70% (2-way), 88.95% (3-way)
Lao et al. [30]	37	Transfer learning	0.739 (c-Index)
Molina-Garcia et al. [31]	93	Neural networks	0.825 (c-Index)
Chaddad et al. [3]	40	Random forest	0.85 (AUC)
Yang et al. [26]	82	Random forest	0.69 (AUC)

4.2 Predictions of Infiltration and Recurrence

Glioblastoma appear to have fairly uniform infiltration, suggested by the edematous (bright FLAIR) imaging characteristics. Infiltration patterns, throughout the edema, are spatially heterogeneous and deeply penetrating. Despite the difficulties in demarcating the infiltrating neoplasm from edema tissue by using conventional qualitative approaches, machine learning methods have shown a substantial promise in identifying margins of infiltrative tissue on preoperative MR images. The demarcation of the infiltrated tissue, which appears to be at high risk of tumor recurrence, may lead to guide extended surgical resections, localized biopsies, and radiation treatment planning.

Akbari et al. [32] demonstrated that advanced pattern analysis and machine learning methods can provide predictive spatial maps of tumor infiltration and the likelihood of early recurrence. Preoperative multi-parametric magnetic resonance images were combined using support vector machine classifier, thereby creating predictive spatial maps of infiltrated peritumoral tissue. This technique produced predictions of early recurrence with a mean area under the curve of 0.84. The method was later extended by Rathore et al. [23, 33, 34] by implementing comprehensive quantitative analysis of distance, texture, statistical, and signal strength measures of the infiltrated and non-infiltrated regions from a large cohort of *de novo* glioblastoma patients using conventional and advanced imaging sequences.

Pathologic-radiologic correlations have also demonstrated significant contribution in assessment of infiltration in peritumoral edema region. Chang et al. [35] registered biopsy sites to the preoperative MRI sequences by using a CNN. They used multimodal imaging features at the biopsy sites to train a network on a cell density counting method applied to pathology images, and found a negative correlation between cellularity and apparent diffusion coefficient and a positive correlation between degree of enhancement and cellularity (Table 2).

Table 2. Summary of the ML based radiomic signatures developed for an assessment of the infiltrating tumor cells of brain tumors

Study	Patients	ML algorithm	Accuracy
Akbari et al. [32]	31	SVM	0.84 (AUC)
Rathore et al. [23]	59	SVM	0.91 (AUC)

4.3 Response Assessment of Brain Tumors

The current standard of care for glioblastoma patients is maximum safe surgical resection, followed by the concomitant and adjuvant chemoradiotherapy. The treatment options generally need to be adjusted in time at different stages of the disease management. An accurate and timely assessment of treatment response is very important and critical in personalized medicine. The currently used method for response assessment is based on Macdonald Criteria, which evaluates treatment response via quantification of the contrast-enhancing regions on MRI scans [36]. This method has limitations because it only considers the contrast-enhanced region of the tumor, and ignores the rest of the regions of the tumors to characterize the treatment response. The treatment response could be either true recurrence/progression or pseudo-progression, which mimics true progression but has characteristics of normal tissue.

There is mounting evidence in the recent past that machine learning when applied on MRI can reveal the characteristics of recurred tissue, and may help distinguishing between pseudo-progression and true recurrence. Automatic machine learning approaches that incorporate multi-parametric MRI features have had success in predicting pseudoprogression [37–39]. In one radiomic study with 304 patients, Abrol et al. [40] identified 100 significant features that were used to build a SVM model. Booth et al. [41] proposed a technique that distinguishes pseudo-progression from true progression by analyzing tumor heterogeneity in T_2-weighted images using topological descriptors of image heterogeneity called Minkowski functionals (MFs). Akbari et al. [42] used multi-parametric MRI to extract extensive features and conduct support vector machine based analysis to distinguish true and pseudo-progression with 83% accuracy.

Deep learning methods have also shown some promise in distinguishing between pseudo-progression and true recurrence. In this context, Jang et al. [43] investigated the potential role of CNN with long short term memory (CNN-LSTM) structure in discrimination of pseudo-progression and true progression. They used conventional images, especially gadolinium-enhanced T1-weighted MRI, in patients with glioblastoma after concurrent chemoradiotherapy (CCRT). CNN was used to learn features from brain MRI and LSTM was used to recognize the spatial sequence of images. Clinical parameters including age, gender, total radiation dose, number of fractions, interval between CCRT and appearance of lesion were also utilized in their study. Their algorithm achieved a moderate predictability with an area under the curve (AUC) of 0.83 (Table 3).

Table 3. Summary of the ML based radiomic signatures developed for an assessment of tumor response (AUC = area under the curve)

Study	Patients	ML algorithm	Accuracy
Jang et al. [43]	19	CNN-LSTM	0.83 (AUC)
Qian et al. [39]	35	SVM	0.92 (AUC)
Hu et al. [37]	31	SVM	0.94 (AUC)
Parekh et al. [38]	24	SVM	0.93 (AUC)
Abrol et al. [40]	304	SVM	90%
Booth et al. [41]	50	SVM	85%
Akbari et al. [42]	23	SVM	83%

4.4   Characterization of Imaging Heterogeneity

Accurate characterization of the heterogeneity of brain tumors is critical not only for better understanding of brain tumors, but also for developing personalized therapies to improve patient outcome, and for facilitating targeted enrollment into clinical trials [44]. Radiomic analysis of mpMRI has been used as a powerful diagnostic method for in vivo characterization of diverse aspects of brain tumors and their micro environment [45, 46]. Itakura et al. [45] identified subtypes of glioblastoma using consensus clustering of various image features from MR images. They identified three clusters – premultifocal, spherical and rim-enhancing. Each cluster mapped to unique set of molecular signaling pathways using pathway activity estimates and also demonstrated differential survival probabilities. Rathore et al. [22, 47, 48] applied advanced imaging analytics and radiomic approaches to multi-parametric MRI of *de novo* glioblastoma patients ($n = 208$ discovery, $n = 53$ replication), and discovered three distinct and reproducible imaging subtypes of glioblastoma, with differential clinical outcome and underlying molecular characteristics, including isocitrate dehydrogenase-1 (*IDH1*), O^6-methylguanine–DNA methyltransferase, epidermal growth factor receptor variant III (*EGFRvIII*), and transcriptomic subtype composition. The subtypes provided risk-stratification substantially beyond that provided by WHO classifications [1] (Table 4).

Table 4. Summary of the MRI based methods developed to determine intrinsic imaging subtypes (homogeneous patients subgroups with in a heterogeneous population) of brain tumors

Study	Patients	ML algorithm	Accuracy
Itakura et al. [45]	37	Consensus clustering	P = 0.004 (Logrank)
Rathore et al. [22]	53	Consensus clustering	P = 0.001 (Logrank)

5   Discussion

Despite the overwhelming growth in the field of machine learning, most of this research did not make to clinical settings. Many studies using machine learning have reported more than 90% success rate in predicting clinical outcome of interest.

Although, the authors have reported cross-validated findings in these studies, the retrospective nature of those analyses with training makes the models prone to overfitting the data, and biases the results towards higher performance. Freezing the models, and applying them on new data, possibly with slightly different imaging protocols would probably degrade that performance. Another technical limitation is the relative paucity of cases used to train those models. There is a need to validate those studies with systematic prospective testing, which has not yet been achieved. The research has proposed relatively small studies, yielding journal publications, but not more than that. Large-scale, prospective validation is needed to increase confidence in those prediction studies. What is needed is not only studies involving multiple centers to make up large patient populations, but also a consensus on the algorithms used. The latter is not easily achieved. One needs a mechanism to compare the algorithms offered by different labs, and a way to build an optimal one based on what's already available. There have been several initiatives to do that, but the next step, which is to determine the optimal solution from the best algorithms has not been taken. Such mechanisms to determine the standard algorithm for a particular task would probably be required in order to generate enough confidence in those computing techniques, so they can be introduced into the clinical arena. Of course, as the technology evolves, the standard is bound to become obsolete with time, and a mechanism for keeping up to date would also be needed.

The analysis and the results of machine learning based studies leads to an important question "what clinical impact can those predictions have?" The management of brain tumor patients involves standard clinical practice, like any medical treatment or procedure. This standard of care is the result of tradition or scientific evidence, for example acquired as a result of clinical trials. In the current regimen of clinical practice, the only way to change standard of care is through clinical trials. So, in order for those machine learning techniques to have a clinical impact is to put them to the clinical trial test. However, in order to initiate such trials utilizing the output of machine learning algorithms in making clinical decisions, one would need to overcome a natural reluctance from neurosurgeons, neuro-oncologists and radiation oncologists. Broadly speaking, such techniques can make changes in surgical or chemoradiation management based on imaging findings, for example, more aggressive surgery or radiation therapy in a non-enhancing area of high probability of recurrence. However, it is probably easier to embark in a clinical trial that tests a new chemotherapeutic agent than to test new patient management on the basis of predictions from a computer algorithm.

6 Conclusion

We have presented a number of methodologies, focused on the topic of machine learning and MRI in Neuro-Radiomics. Radiomics depends on huge amounts of clinical data (such as MRI scans) to validate it's machine-learning approaches and development of predictors. We propose that the data should be carefully curated in consultation with the radiologists to develop reliable predictors of clinical outcome of interest. Despite the growing use of these algorithms in academic and research settings, there are major hurdles to the efficient deployment of these sophisticated algorithms in a clinical setting and their seamless integration into routine clinical workflows. While the radiologists' opinions will still be considered the gold standard, it would be crucial for the radiologists to learn, understand and appropriately use these powerful tools as they become more available in near future.

References

1. Louis, D.N., et al.: The 2016 world health organization classification of tumors of the central nervous system: a summary. Acta Neuropathol. **131**, 803–820 (2016)
2. Hardee, M.E., Zagzag, D.: Mechanisms of glioma-associated neovascularization. Am. J. Pathol. **181**, 1126–1141 (2012)
3. Chaddad, A., et al.: Prediction of survival with multi-scale radiomic analysis in glioblastoma patients. Med. Biol. Eng. Comput. **56**, 2287–2300 (2018)
4. Asai, A., et al.: Subacute brain atrophy after radiation therapy for malignant brain tumor. Cancer **63**, 1962–1974 (1989)
5. Shukla, G., et al.: Advanced magnetic resonance imaging in glioblastoma: a review. Chin. Clin. Oncol. **6**, 40 (2017)
6. Villanueva-Meyer, J.E., et al.: Current clinical brain tumor imaging. Neurosurgery **81**, 397–415 (2017)
7. Lambin, P., et al.: Radiomics: extracting more information from medical images using advanced feature analysis. Eur. J. Cancer **48**, 441–446 (2012)
8. Kumar, V., et al.: Radiomics: the process and the challenges. Magn. Reson. Imaging **30**, 1234–1248 (2012)
9. Rathore, S., et al.: Radiopathomics: integration of radiographic and histologic characteristics for prognostication in glioblastoma. Soc. Neuro-Oncol. (2019)
10. Gillies, R.J., et al.: Radiomics: images are more than pictures, they are data. Radiology **278**, 563–577 (2015)
11. Zhou, M., et al.: Radiomics in brain tumor: image assessment, quantitative feature descriptors, and machine-learning approaches. Am. J. Neuroradiol. **39**, 208–216 (2018)
12. Yushkevich, P.A., et al.: User-guided 3D active contour segmentation of anatomical structures: significantly improved efficiency and reliability. NeuroImage **31**, 1116–1128 (2006)
13. Gaonkar, B., et al.: Automated segmentation of brain lesions by combining intensity and spatial information. In: IEEE International Symposium on Biomedical Imaging (ISBI), pp. 93–96 (2010)
14. Lian, Y., Song, Z.: Automated brain tumor segmentation in magnetic resonance imaging based on sliding-window technique and symmetry analysis. Chin. Med. J. **127**, 462–468 (2014)
15. Lu, S., et al.: Peritumoral diffusion tensor imaging of high-grade gliomas and metastatic brain tumors. Am. J. Neuroradiol. **24**, 937–941 (2003)
16. Wintermark, M., et al.: Comparative overview of brain perfusion imaging techniques. J. Neuroradiol. **32**, 294–314 (2005)
17. Tykocinski, E.S., et al.: Use of magnetic perfusion-weighted imaging to determine epidermal growth factor receptor variant III expression in glioblastoma. Neurooncol. **14**, 613–623 (2012)
18. Fedorov, A., et al.: 3D slicer as an image computing platform for the Quantitative Imaging Network. Magn. Reson. Imaging **30**, 1323–1341 (2012)

19. van Griethuysen, J.J.M., et al.: Computational radiomics system to decode the radiographic phenotype. Cancer Res. **77**, e104–e107 (2017)
20. Davatzikos, C., et al.: Cancer imaging phenomics toolkit: quantitative imaging analytics for precision diagnostics and predictive modeling of clinical outcome. J. Med. Imaging: Spec. Sect. Quant. Imaging Methods Transl. Dev. – Honoring Mem. Dr. Larry Clarke **5**, 011018 (2018)
21. Macyszyn, L., et al.: Imaging patterns predict patient survival and molecular subtype in glioblastoma via machine learning techniques. Neuro Oncol. **18**, 417–425 (2016)
22. Rathore, S., et al.: Radiomic MRI signature reveals three distinct subtypes of glioblastoma with different clinical and molecular characteristics, offering prognostic value beyond IDH1. Nat. Sci. Rep. **8**, 5087 (2018)
23. Rathore, S., et al.: Radiomic signature of infiltration in peritumoral edema predicts subsequent recurrence in glioblastoma: implications for personalized radiotherapy planning. J. Med. Imaging (Bellingham) **5**, 021219 (2018)
24. Akbari, H., et al.: Quantitative radiomics and machine learning to distinguish true progression from pseudoprogression in patients with GBM. In: 56th Annual Meeting, American Society of NeuroRadiology (ASNR) (2018)
25. Arbabshirani, M.R., et al.: Single subject prediction of brain disorders in neuroimaging: promises and pitfalls. NeuroImage **145**, 137–165 (2017)
26. Yang, D., et al.: Evaluation of tumor-derived MRI-texture features for discrimination of molecular subtypes and prediction of 12-month survival status in glioblastoma. Med. Phys. **42**, 6725–6735 (2015)
27. Rathore, S., et al.: Quantitative imaging predictors of overallsurvival in glioblastoma patients robust in the presence of inter-scanner variations. Soc. Neuro-Oncol. **20**(Suppl. 6), vi184 (2018)
28. Chato, L., Latifi, S.: Machine learning and deep learning techniques to predict overall survival of brain tumor patients using MRI images. In: 2017 IEEE 17th International Conference on Bioinformatics and Bioengineering (BIBE), pp. 9–14 (2017)
29. Krizhevsky, A., et al.: ImageNet Classification with Deep Convolutional Neural Networks. NIPS (2012)
30. Lao, J., et al.: A deep learning-based radiomics model for prediction of survival in glioblastoma multiforme. Sci. Rep. **7**, 10353 (2017)
31. Molina-García, D., et al.: Prognostic models based on imaging findings in glioblastoma: human versus machine. Sci. Rep. **9**, 5982 (2019)
32. Akbari, H., et al.: Imaging surrogates of infiltration obtained via multiparametric imaging pattern analysis predict subsequent location of recurrence of glioblastoma. Neurosurgery **78**, 572–580 (2016)
33. Rathore, S., et al.: Technical note: a radiomic signature of infiltration in peritumoral edema predicts subsequent recurrence in glioblastoma. In: Medical Imaging 2018: Image-Guided Procedures, Robotic Interventions, and Modeling, vol. 10576, p. 105760O (2018)
34. Sloan, A.E., et al.: Radiomics-based identification of peritumoral infiltration in de novo glioblastoma imaging presents targets amenable for potential targeted extended resection: a neurosurgical survey. J. Clin. Oncol. **37**, e13573 (2019)
35. Chang, P.D., et al.: A multiparametric model for mapping cellularity in glioblastoma using radiographically localized biopsies. AJNR Am. J. Neuroradiol. **38**, 890–898 (2017)
36. Macdonald, D.R., et al.: Response criteria for phase II studies of supratentorial malignant glioma. J. Clin. Oncol. **8**, 1277–1280 (1990)
37. Hu, X., et al.: Support vector machine multiparametric MRI identification of pseudoprogression from tumor recurrence in patients with resected glioblastoma. J. Magn. Reson. Imaging: JMRI **33**, 296–305 (2011)

38. Parekh, V., et al.: Multiparametric Deep Learning and Radiomics for Tumor Grading and Treatment Response Assessment of Brain Cancer: Preliminary Results (2019)
39. Qian, X., et al.: Stratification of pseudoprogression and true progression of glioblastoma multiform based on longitudinal diffusion tensor imaging without segmentation. Med. Phys. **43**, 5889 (2016)
40. Abrol, S., et al.: Radiomic analysis of pseudo-progression compared to true progression in glioblastoma patients: a large-scale multi-institutional study. J. Clin. Oncol. **35**, 2015 (2017)
41. Booth, T.C., et al.: Analysis of heterogeneity in T2-weighted MR images can differentiate pseudoprogression from progression in glioblastoma. PLoS ONE **12**, e0176528 (2017)
42. Akbari, H., et al.: Quantitative image analysis and machine learning techniques for distinguishing true progression from pseudoprogression in patients with glioblastoma. J. Neuro-Oncol. **20**, vi191–vi192 (2018)
43. Jang, B.-S., et al.: Prediction of pseudoprogression versus progression using machine learning algorithm in glioblastoma. Sci. Rep. **8**, 12516 (2018)
44. Davatzikos, C., et al.: Precision diagnostics based on machine learning-derived imaging signatures. Magn. Reson. Imaging **64**, 49–61 (2019)
45. Itakura, H., et al.: Magnetic resonance image features identify glioblastoma phenotypic subtypes with distinct molecular pathway activities. Sci. Transl. Med. **7**, 303ra138–303ra138 (2015)
46. Kickingereder, P., et al.: Radiomic profiling of glioblastoma: identifying an imaging predictor of patient survival with improved performance over established clinical and radiologic risk models. Radiology **280**, 880–889 (2016)
47. Rathore, S., et al.: Radiologic subtypes of glioblastoma calculated via multi-parametric imaging signatures reveal complementary information to current Who classification. Neuro-Oncol. **19**, vi155–vi156 (2017)
48. Rathore, S., et al.: Imaging pattern analysis reveals three distinct phenotypic subtypes of GBM with different survival rates. Neuro-Oncol. 18, vi128 (2016)

Opportunities and Advances in Radiomics and Radiogenomics in Neuro-Oncology

Kaustav Bera, Niha Beig, and Pallavi Tiwari[✉]

Case Western Reserve University, Cleveland, OH, USA
pallavi.tiwari@case.edu

Abstract. Neuro-oncology broadly encompasses life threatening malignancies of the brain and spinal cord including both primary as well as lesions metastasizing to the central nervous system. The biggest clinical challenge in the field currently is to be able to design personalized treatment management solutions in patients based on *apriori* knowledge of their survival outcome or response to conventional or experimental treatments. *Radiomics* or the quantitative extraction of subvisual data from conventional radiographic imaging and *radiogenomics*, statistically correlating radiomic features with point-mutations and next generation sequencing data, have recently emerged as unique mechanisms to offer insights into answering some of these clinically relevant questions related to diagnosis, classification, prognosis as well as assessing treatment response. In this review, we provide an overview of the framework for radiomic and radiogenomic approaches in neuro-oncology, including a brief description of the techniques commonly employed. Further, we will provide a review of some of the existing applications of radiomics and radiogenomics in neuro-oncology for tumor classification, survival prognosis, predicting response to therapies, as well as distinguishing benign post-treatment changes from tumor recurrence, using routine MRI scans. While highly promising, the clinical acceptance of radiomics and radiogenomics techniques will largely hinge on their resilience to non-standardization across imaging protocols, as well as in their ability to demonstrate reproducibility across large multi-institutional cohorts.

Keywords: Radiomics · Radiogenomics · Neuro-oncology · Treatment response · Prognosis

1 Introduction

The field of neuro-oncology includes potentially life-threatening primary as well as metastatic malignant tumors of the brain and spinal cord. Despite the relative rarity of malignancy in the brain as compared to other parts of the body, brain tumors have exceptionally higher mortality and morbidity [1, 2]. Among the most malignant tumors, brainstem gliomas, glioblastomas (GB) and anaplastic astrocytomas have the worst outcome [3, 4]. In addition to precise microsurgical techniques, advances in multiplanar imaging has been one of the cornerstones

© Springer Nature Switzerland AG 2020
H. Mohy-ud-Din and S. Rathore (Eds.): RNO-AI 2019, LNCS 11991, pp. 12–23, 2020.
https://doi.org/10.1007/978-3-030-40124-5_2

in neuro-oncology. Diagnosis, prognosis, and treatment response assessment in brain tumors is currently investigated using multiparametric MRI (mpMRI), typically including T1 weighted imaging both before (T1w) and after administration of gadolinium based contrast agent (T1c), T2 weighted imaging (T2w), and T2w-Fluid attenuation recovery (FLAIR) sequences. T1w imaging especially after gadolinium based contrast enhancement is useful for elucidating a brain tumor, with large portions of the tumoral region (or the entire tumor) typically enhancing in comparison to the brain parenchyma. Meanwhile T2w and FLAIR imaging are frequently used for heterogeneously enhancing tumors like gliomas especially GBs which often have a necrotic core of dead tissue as well as peritumoral edema or inflammation, in order to differentiate these regions from the actual tumor. In addition to qualitative (visual) evaluation of these imaging modalities by expert neuro-radiologists, they also harbor enormous amounts of interpretable quantitative information which may not be appreciable by the naked eye [5]. Recent advances in computational processing power and high-throughput algorithmic development has led to the development of novel quantitative image analysis methods radiomics and radiogenomics which leverage routine mpMRI imaging to interrogate the lesion environment towards early detection, diagnosis, prognosis as well as treatment response in neuro-oncology.

1.1 What Is Radiomics?

Radiomics, a term coined by Lambin et al. [6] is the high-throughput extraction of quantitative information from routine radiological images (X-Rays, CT, MRI, PET) for defining the textural and morphological characteristics of the disease. Radiomics has been widely used across multiple cancer subtypes including brain [7–14], lung [15–19], colorectal [20,21], breast [22,23], prostate [24,25] among others for disease diagnosis, prognosis, outcome prediction as well as measuring early treatment changes or gauging response to treatment. Radiomics comprehensively and quantitatively characterizes pixel-wise tumor characteristics including (a) shape features which provide quantitative measures of how regular or irregular the tumor boundaries change based on their 3D topology, (b) semantic or qualitative features which includes radiologist derived assessments of the tumor including spiculations, size of the tumor along several axes, (c) intra-tumoral heterogeneity measures including gray-level features, which investigate pixel level differences in the texture of the tumor in order to characterize how heterogeneous a tumor is, as well as (d) peri-lesional features including deformation features which capture the impact of tumor related mass effect in the tumor micro-environment (TME) [15,22,23].

1.2 What Is Radiogenomics?

Gene mutations are one of the hallmarks of cancer with single-gene or multi-gene mutations determining the aggressiveness of the tumor, its growth patterns as well as its response to treatment. In neuro-oncology, in spite of multiple studies demonstrating that driver mutations and multi-gene expression as well as transcriptomic pathways are prognostic across a variety of brain tumors, there currently exists no validated molecular biomarker for prognosis or assessing treatment response. The most recent advancement in this field has been the 2016 update on the WHO classification of diffuse gliomas which for the first time, now includes specific prognostic point mutations as well as chromosomal alterations [26]. In GBs, multiple studies have demonstrated prognostic significance of driver mutations especially of IDH1, EGFRviii, PTEN, 1p/19q, BRAF as well as MGMT methylation among others including multi-gene expression pathways [27–29].

These advances have led to a surge of interest in the field of radiogenomics [30,31] wherein quantitative imaging or the radiomic phenotype of the tumor can be correlated with the underlying genetic profile of the tumor, including point mutations, signaling and pathways of biological significance. Radiogenomics, by providing an imaging phenotype for the entire tumor corresponding to a particular genotype, might play a role in circumventing the issues of intratumoral heterogeneity in biopsy sampling, by spatially mapping the entirety of the tumor, instead of the analysis being dependent on a particular biopsy location [32].

2   The Radiomics/Radiogenomics Pipeline

2.1   Pre-processing and Segmentation

Pre-processing involves multi-protocol registration, skull stripping, and intensity standardization to account for variations in MRI scans across different manufacturers, magnetic strengths, and slice thicknesses. Following pre-processing, the next important step is the accurate and reliable delineation of the tumor ROI. While manual segmentations ensure high accuracy, and are currently the gold standard for radiomic analysis [18], there are several automatized segmentation approaches using deep learning architectures including U-Net, Conv-Net, Transfer Learning, and Deep Hourglass approaches as well as semi-automatic seeding-based algorithms that have found popularity [33]. These sophisticated automated tools annotate the tumor sub-compartments including enhancement, non-enhancement, necrosis as well as specific sub compartments in the immediate periphery of the tumor.

2.2   Quantitative Features Extraction

Following segmentation, the next step is to extract quantitative features that can capture multiple sub-visual morphometric and textural features of tumor heterogeneity on routine MRI scans. The most commonly employed radiomic features that are utilized in brain tumor analysis include:

- **Semantic features** - In a large-scale TCGA study of brain tumors (low grade gliomas and GBs), neuro-radiologists defined features based on the visual phenotypic characterization of the tumors. Distinctive features were based on

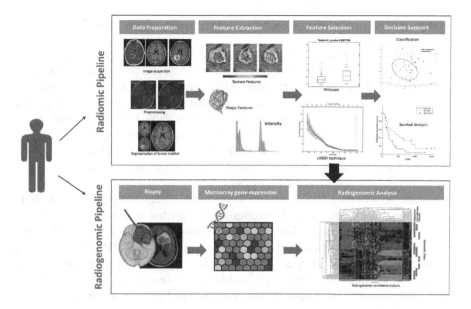

Fig. 1. Overall workflow of radiomic and radiogenomic pipeline

the four commonly analyzed tumor sub-compartments (non-enhanced tumor, enhancing tumor, necrosis and edema), and included features such as location of the lesion, morphology, major axis length, minor axis length margin, and lesion vicinity. This comprehensive radiologist defined feature set is now known as VASARI (Visually AcceSAble Rembrandt Images) [34].

- **Shape based features** - Irregular and aggressive tumor infiltration can induce surface and shape changes in the tumor and peritumoral regions. Some of the global shape features include the major and minor axis, elongation (ratio between major and minor axes of the ROI) of the shape of the segmented compartments. Local surface features capture characteristics such as curvature which identifies flat areas of surface from highly curved ones, sharpness which measures how sharp the curvature is, with highly curved masses exhibiting sharper curvatures, as well as shape index which characterizes the shape topology of the tumor [9,28].

- **Texture features** - The most commonly used texture features [6,8,18] include: (1) Gray-level co-occurrence matrix (GLCM) which involve capturing variations in gray-level image characteristics via second order intensity statistics (e.g. angular second moment, contrast, and differential entropy), (2) Gray-level run length matrix (GLRLM), which in comparison with GLCM features, looks to analyze the pixel runs instead of pairs of pixels. A pixel run includes the number of pixels of a specific gray value that are in a right direction, in the right sequence. While the rows of the matrix still represent gray levels, the columns represent run lengths, and (3) Laws features which define various texture parameters including spot, edge, ripple and level surfaces present within the tumor.

- **Deformation features** - These are relatively new feature descriptors [35] that seek to measure tissue deformation present in the brain parenchyma due to mass effect in brain tumors. MRI scans of the diseased patient are non-rigidly registered to corresponding healthy, age-matched, and/or gender-matched imaging atlases. The resulting deformation field (represented as a displacement vector at every voxel location) is obtained through a combination of forward as well as inverse mapping between the patient 3D volume and the reference atlas. The per-voxel deformation measurements are then used as radiomic features for analysis.

- **Wavelet features** - These are features that utilize different wavelengths, amplitudes and frequencies to recognize attributes across a wide range of scales within the image. For instance, Gabor wavelet features [36] captures image gradients across varying frequencies and wavelengths. Similarly, a recently introduced radiomics wavelet descriptor, Co-occurrence of local anisotropic gradient orientations (CoLIAGe) [37] builds on the existing Gabor filter and captures the apparent disorder in gradient orientations down to a pixel basis within an image.

2.3   Classifier Construction and Analysis

Constructing the classifier first involves trimming the feature set by selecting the most discriminative features using a feature selection scheme, in order to reduce the curse of dimensionality [38]. The next step involves creation of a machine learning model that can predict outcome or evaluate response, by categorizing and classifying various datasets according to defined labels. Broadly categorizing, classifiers can be divided into supervised and unsupervised approaches. While supervised methods utilize a pre-defined set of known labels to represent the outcomes of interest in the images analyzed, an unsupervised approach such as clustering can be employed when the target labels are unknown [18].

2.4   Radiogenomic Analysis

Genomic analysis involves analyzing the fresh frozen paraffin embedded (FFPE) sample or the tissue microarray (TMA) sample from a stereotactic brain biopsy from within the tumor. In the simplest form, this might involve detecting single-gene mutations, for instance EGFR amplification, MGMT methylation by analyzing the proteins through immunohistochemistry (IHC) analysis, to multi-gene expression analysis through next generation sequencing which might involve mRNA sequencing, miRNA sequencing, whole exome sequencing, whole genome sequencing among others. Radiogenomic analysis could involve correlating single gene mutations directly with individual radiomic features or the radiomics derived risk model, as well as correlating with gene sets responsible for downstream pathways of biological significance obtained using gene set enrichment analysis (GSEA). With available gene ontologies of common biological pathways, this enables a gene ontology analysis which can be used to derive pathways which are over-expressed/under expressed in radiomic derived risk groups.

3 Applications in Neuro-Oncology

3.1 Classification and Grading of Brain Tumors

Multiple studies [39–41] have recently explored radiomic and radiogenomic differences between the WHO defined classes of gliomas. For instance, in $N = 214$ (106 GBs and 108 low grade gliomas) as training and an independent validation set of $N = 70$, Lu et al. [41] showed that a radiomics model achieved an area under the receiver operating curve (AUC) of 0.922 and 0.975 and accuracies between 87.7% and 96.1% to predict the IDH1 and 1p/19q status. In the independent validation set, the radiomics model had an accuracy of 81.8% accuracy in classifying glioma into one of the five molecular classed. Meanwhile Cho et al. [40] used $N = 285$ gliomas and applied radiomic methods across T1w, T1 contrast enhanced, T2 and FLAIR MRI modalities to build three machine learning classifiers to separate low grade from high grade gliomas. The authors obtained AUCs of 0.94 for training cohorts and 0.9030 (logistic regression 0.90, support vector machine 0.88, and random forest 0.92) for testing on five-fold cross validation. Yang et al. similarly used T1w and FLAIR modalities from $N = 82$ patients with GBM and used radiomic textural features on MRI to differentiate between the four molecular subtypes of GBM, with AUCs ranging from 0.72–0.82 with T1 Haralick, fractal and oriented gradient features being significant. Interestingly, Kickingereder et al. [39] showed that there was no specific locational differences for the different GBM subtypes, while also finding correlations between radiomic features and some molecular characteristics including EGFR amplification and RTKII GB subgroup.

3.2 Survival Risk Stratification in Brain Tumors

Prasanna et al. [7] used $N = 65$ GB patients and used MRI radiomic features of T1w, T2w and FLAIR sequences from the tumor and peritumoal regions to predict short term (<7 months) versus long term survival (>18 months). Using threefold cross validation, the authors found that peritumoral MRI features corresponding to heterogeneous intensity and textural patterns could predict survival ($p = 1.47 \times 10-5$). Tixier et al. [42], showed that in $N = 159$ patients with untreated GBM, a radiomic model could be used to accurately predict overall survival (OS). Additionally, they found that combining MGMT methylation status with the radiomic descriptors allowed the identification of a group of patients with good prognosis (survival probability of 0.61 after 43 months; $p = 0.0005$). Meanwhile Bae et al. [43] used a random forest machine learning classifier fed with radiomic textural features from mpMRI and found that on a validation set of $N = 54$ patients, the radiomic model could stratify patients into a high- and low-risk group based on OS (HR = 2.58; 95% CI = 1.07–6.26)

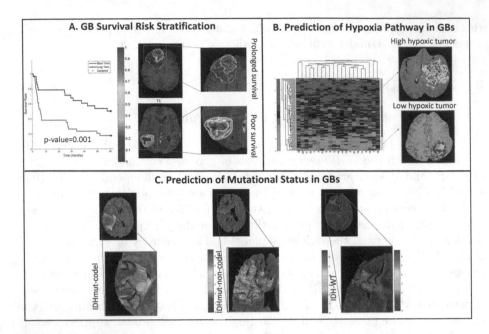

Fig. 2. Radiomics and radiogenomics are used in various applications for Neuro-Oncology. A. Survival in Glioblastomas (GBs) can be predicted using radiomics. B. Radiogenomic analysis of the hypoxia signaling pathway in GBs can non-invasively predict the extent of hypoxia using radiomic features. C. Radiogenomic analysis of GBs can also be used to predict point mutations such as IDH mutations and 1p19q co-deletions status.

and Progression-free survival (PFS) (HR = 3.24; 95% CI = 1.23–8.50). The most discriminative radiomic features included tumor shape, and first order features from the enhancing, edema and necrotic regions.

3.3 Characterizing Post-treatment Changes in Brain Malignancies

Radiomics could serve as a potent tool in accurately characterizing the post-treatment environment in brain tumors. Current response assessment methods in neuro-oncology rely on conventional imaging modalities that measure the maximum diameter of a lesion, using the criteria developed by the Response Assessment in Neuro-Oncology group (RANO criteria) [44]. Conventional RANO criteria are unfortunately limited in distinguishing between benign confounders due to post-radiation effects (i.e. psuedo-progression and radiation necrosis) and true tumor recurrence. Tiwari et al. [8] in a feasibility study using N = 43 patients for training and N = 15 as an independent validation cohort with both primary and metastatic brain malignancies, a radiomic textural method could accurately distinguish radiation necrosis from true tumor recurrence on post chemo-radiotherapy MRIs. The authors found that in the independent validation cohort, the SVM based machine learning classifier identified 12 out of the

15 studies correctly, while 2 experienced neuro-radiologists accurately diagnosed 7 and 8 out of the 15 cases respectively. Meanwhile Ismail et al. [9] demonstrated that radiomic shape features of the TME or tumor habitat could effectively distinguish between pseudo-progression and tumor recurrence on post-treatment mpMRI of patients with GB. On an independent test set of $N = 46$, the authors showed that leveraging radiomic features capturing the total curvature of the enhancing tumor and curvedness of the T2w/FLAIR TME region, the radiomic method had an accuracy of 90.2% in distinguishing pseudo-progression (radiation necrosis) from true tumor recurrence. Using voxel-based intensity radiomic descriptors fed into a SVM machine learning classifier, Hu et al. [10] had an AUC of 0.94 in differentiating between pseudoprogression and true recurrence in $N = 31$ GB patients using post chemoradiation mpMRI. Kickingereder et al. [13] used $N = 172$ GB patients treated with bevacizumab, an anti-angiogenic agent and used pre-treatment mpMRI to extract 4842 quantitative radiomic features. In $N = 60$ as the validation set, using 72 most discriminative radiomic features, the authors could distinguish responders vs. non-responders to bevacizumab based on OS (HR $= 2.60$, p $= 0.001$) and PFS (HR $= 1.85$, p $= 0.030$).

3.4   Radiogenomic Analysis of Brain Tumors

Several recent studies have used radiogenomic analysis to define a genetic signature for radiomic based imaging phenotypes, as a surrogate for single-gene and multi-gene mutation analysis [11,45–47]. Beig et al. [14] used a MRI-based radiomic model as a non-invasive surrogate to characterize hypoxia in GB, which has been shown to be linked with poor survival and chemo refractoriness. Using $N = 85$ GB patients to train the model on a Hypoxia enrichment score (HES) comprising 21 hypoxia-associated genes, a radiomic model correlated with HES was used to stratify GB patients based on their survival. On an independent validation set of $N = 30$ patients, the radiomic features which were strongly associated with HES, could also distinguish short term survivors (OS < 7 months) from long term survivors (OS > 16 months) (p $= 0.003$). Gutman et al. [48] on $N = 76$ GB patients, used volumetric radiomic features from T1c and FLAIR to predict the mutational status of the tumors. The authors found that TP53, RB1, NF1, EGFR, and PDGFRA mutations could be significantly predicted by at least one significant radiomic feature. Geveart et al. [49] meanwhile correlated radiomic features with multi-gene mRNA sequencing data available for $N = 55$ patients from TCGA/TCIA. The authors found seven radiomic features to be significantly correlated with molecular subgroups (p < 0.05) and constructed a radiogenomics map to link the radiomic features with gene expression information related to biological processes. Hu et al. [50] used radiogenomic analysis to show the existence of intra-tumoral heterogeneity (ITH). To investigate spatial ITH, the authors collected 48 biopsies from different spatial locations in 13 tumors and identified radiomic associations for 6 driver mutations (EGFR, PDGFRA, PTEN, CDKN2A, RB1, and TP53). They co-registered biopsy locations with radiomic texture maps to correlate regional genetic status with spatially matched imaging measurements. Their radiogenomic model for 4 driver mutations (EGFR, RB1, CDKN2A, and PTEN) was more accurate in the biopsies from the areas outside the enhancing core of the tumor (n = 16) as compared to the enhancing region (n = 32).

4   Limitations

A significant limitation of existing radiomics and radiogenomic methodologies is that they are currently developed on small datasets, which are often retrospective and from a single institution. Prospective, multi-institutional validation of these tools, preferably in a controlled, homogenous clinical trial setting will be a substantial step towards validating the efficacy and generalizability of these techniques. Another significant limitation of radiomics techniques is the relative lack of understanding of the biological underpinnings driving the expression of these radiomic features. While radio-genomics to some extent has attempted to provide biological associations of radiomic features with molecular processes, deep learning approaches are still largely considered black-box. Clinical translation of these techniques would require unified efforts from all stakeholders including physicians to be involved in development of these tools from conception to deployment, as well as computer scientists to ensure that the tools being developed are tailored and aligned to the needs of the clinical end-users.

5   Future Scope

Radiomics and radiogenomics have made remarkable progress in the last five years itself to provide significant value in the field of oncology at large, as well as neuro-oncology, towards diagnosing, outcome prediction, as well as evaluating response to both conventional and experimental treatments. Interestingly, the radiomics/radiogenomics field is now arriving at an inflection point where the next logical step would be to move these techniques from the realm of research onto clinical practice. Additionally, the tools will need to be developed such that they are easily accessible to clinical end-users including neuro-oncologists, neuro-radiologists, surgeons and physicians, preferably without disrupting their clinical workflow.

References

1. Rouse, C., Gittleman, H., Ostrom, Q.T., Kruchko, C., Barnholtz-Sloan, J.S.: Years of potential life lost for brain and CNS tumors relative to other cancers in adults in the United States, 2010. Neuro Oncol. **18**(1), 70–77 (2016)
2. Wrensch, M., Minn, Y., Chew, T., Bondy, M., Berger, M.S.: Epidemiology of primary brain tumors: current concepts and review of the literature. Neuro-oncology **4**(4), 278–299 (2002)
3. DeAngelis, L.M.: Brain tumors. N. Engl. J. Med. **344**(2), 114–123 (2001)
4. Fisher, J.L., Schwartzbaum, J.A., Wrensch, M., Wiemels, J.L.: Epidemiology of brain tumors. Neurol. Clin. **25**(4), 867–890 (2007)

5. Gillies, R.J., Kinahan, P.E., Hricak, H.: Radiomics: images are more than pictures, they are data. Radiology **278**(2), 563–577 (2015)
6. Lambin, P., et al.: Radiomics: extracting more information from medical images using advanced feature analysis. Eur. J. Cancer **48**(4), 441–446 (2012)
7. Prasanna, P., Patel, J., Partovi, S., Madabhushi, A., Tiwari, P.: Radiomic features from the peritumoral brain parenchyma on treatment-naive multi-parametric MR imaging predict long versus short-term survival in glioblastoma multiforme: pre-liminary findings. Eur. Radiol. **27**, 4188–4197 (2016)
8. Tiwari, P., et al.: Computer-extracted texture features to distinguish cerebral radionecrosis from recurrent brain tumors on multiparametric MRI: a feasibility study. Am. J. Neuroradiol. **37**(12), 2231–2236 (2016)
9. Ismail, M., et al.: Shape features of the lesion habitat to differentiate brain tumor progression from pseudoprogression on routine multiparametric MRI: a multisite study. Am. J. Neuroradiol. **39**(12), 2187–2193 (2018)
10. Hu, X., Wong, K.K., Young, G.S., Guo, L., Wong, S.T.: Support vector machine multiparametric MRI identification of pseudoprogression from tumor recurrence in patients with resected glioblastoma. J. Magn. Reson. Imaging **33**(2), 296–305 (2011)
11. Kickingereder, P., et al.: Large-scale radiomic profiling of recurrent glioblastoma identifies an imaging predictor for stratifying anti-angiogenic treatment response. Clin. Cancer Res. **22**(23), 5765–5771 (2016)
12. Kickingereder, P., et al.: Radiomic profiling of glioblastoma: identifying an imaging predictor of patient survival with improved performance over established clinical and radiologic risk models. Radiology **280**(3), 880–889 (2016)
13. Rathore, S., et al.: Radiomic MRI signature reveals three distinct subtypes of glioblastoma with different clinical and molecular characteristics, offering prognostic value beyond IDH1. Sci. Rep. **8**(1), 1–12 (2018)
14. Beig, N., et al.: Radiogenomic analysis of hypoxia pathway is predictive of overall survival in Glioblastoma. Sci. Rep. **8**(1), 7 (2018)
15. Beig, N., et al.: Perinodular and intranodular radiomic features on lung CT images distinguish adenocarcinomas from granulomas. Radiology **290**(3), 783–792 (2018)
16. Khorrami, M., et al.: Combination of peri- and intratumoral radiomic features on baseline CT scans predicts response to chemotherapy in lung adenocarcinoma. Radiol.: Artif. Intell. **1**(2), 180012 (2019)
17. Khorrami, M., et al.: Predicting pathologic response to neoadjuvant chemoradi-ation in resectable stage III non-small cell lung cancer patients using computed tomography radiomic features. Lung Cancer **1**(135), 1–9 (2019)
18. Bera, K., Velcheti, V., Madabhushi, A.: Novel quantitative imaging for predicting response to therapy: techniques and clinical applications. Am. Soc. Clin. Oncol. Educ. Book **38**, 1008–1018 (2018)
19. Thawani, R., et al.: Radiomics and radiogenomics in lung cancer: a review for the clinician. Lung Cancer **115**, 34–41 (2018)
20. Antunes, J., et al.: Coregistration of preoperative MRI with ex vivo mesorectal pathology specimens to spatially map post-treatment changes in rectal cancer onto in vivo imaging. Acad. Radiol. **25**, 833–841 (2018)
21. Antunes, J., Prasanna, P., Madabhushi, A., Tiwari, P., Viswanath, S.: RADIomic spatial TexturAl descripTor (RADISTAT): characterizing intra-tumoral hetero-geneity for response and outcome prediction. In: Descoteaux, M., Maier-Hein, L., Franz, A., Jannin, P., Collins, D.L., Duchesne, S. (eds.) MICCAI 2017. LNCS, vol. 10434, pp. 468–476. Springer, Cham (2017). https://doi.org/10.1007/978-3-319-66185-8_53

22. Barbur, I., et al.: Automated segmentation and radiomic characterization of visceral fat on bowel MRIs for Crohns disease. In: Medical Imaging 2018: Image-Guided Procedures, Robotic Interventions, and Modeling. International Society for Optics and Photonics (2018)

23. Huang, X., et al.: CT-based radiomics signature to discriminate high-grade from low-grade colorectal adenocarcinoma. Acad. Radiol. **25**(10), 1285–1297 (2018)

24. Liu, Z., et al.: Radiomics analysis for evaluation of pathological complete response to neoadjuvant chemoradiotherapy in locally advanced rectal cancer. Clin. Cancer Res. **23**(23), 7253–7262 (2017)

25. Braman, N.M., et al.: Intratumoral and peritumoral radiomics for the pretreatment prediction of pathological complete response to neoadjuvant chemotherapy based on breast DCE-MRI. Breast Cancer Res. **19**(1), 57 (2017)

26. Braman, N., et al.: Association of peritumoral radiomics with tumor biology and pathologic response to preoperative targeted therapy for HER2 (ERBB2) positive breast cancer. JAMA Netw. Open. **2**(4), e192561 (2019)

27. Ginsburg, S.B., et al.: Radiomic features for prostate cancer detection on MRI differ between the transition and peripheral zones: Preliminary findings from a multi-institutional study: radiomic Features for Prostate Cancer Detection on MRI. J. Magn. Reson. Imag. **46**(1), 184–193 (2017)

28. Ghose, S., et al.: Prostate shapes on pre-treatment MRI between prostate cancer patients who do and do not undergo biochemical recurrence are different: preliminary findings. Sci. Rep. **7**(1), 1–8 (2017)

29. Louis, D.N., et al.: The 2016 world health organization classification of tumors of the central nervous system: a summary. Acta Neuropathol. **131**(6), 803–820 (2016)

30. Liang, Y., et al.: Gene expression profiling reveals molecularly and clinically distinct subtypes of glioblastoma multiforme. Proc. Natl. Acad. Sci. **102**(16), 5814–5819 (2005)

31. Rich, J.N., et al.: Gene expression profiling and genetic markers in glioblastoma survival. Cancer Res. **65**(10), 4051–4058 (2005)

32. Ellingson, B.M.: Radiogenomics and imaging phenotypes in glioblastoma: novel observations and correlation with molecular characteristics. Curr. Neurol. Neurosci. Rep. **15**(1), 506 (2015)

33. Hesamian, M.H., Jia, W., He, X., Kennedy, P.: Deep learning techniques for medical image segmentation: achievements and challenges. J. Digit. Imaging **32**(4), 582–596 (2019)

34. VASARI Research Project - The Cancer Imaging Archive (TCIA) Public Access - Cancer Imaging Archive. https://wiki.cancerimagingarchive.net/display/Public/VASARI+Research+Project

35. Prasanna, P., et al.: Radiographic-deformation and textural heterogeneity (r-DepTH): an integrated descriptor for brain tumor prognosis. In: Descoteaux, M., Maier-Hein, L., Franz, A., Jannin, P., Collins, D.L., Duchesne, S. (eds.) MICCAI 2017. LNCS, vol. 10434, pp. 459–467. Springer, Cham (2017). https://doi.org/10.1007/978-3-319-66185-8_52

36. Marelja, S.: Mathematical description of the responses of simple cortical cells. JOSA **70**(11), 1297–1300 (1980)

37. Prasanna, P., Tiwari, P., Madabhushi, A.: Co-occurrence of local anisotropic gradient orientations (CoLlAGe): a new radiomics descriptor. Sci. Rep. **6**(1), 37241 (2016)

38. Friedman, J.H.: On bias, variance, 0/1loss, and the curse-of-dimensionality. Data Min. Knowl. Discov. **1**(1), 55–77 (1997)

39. Kickingereder, P., et al.: Radiogenomics of glioblastoma: machine learningbased classification of molecular characteristics by using multiparametric and multiregional MR imaging features. Radiology **281**(3), 907–918 (2016)
40. Cho, H., Lee, S., Kim, J., Park, H.: Classification of the glioma grading using radiomics analysis. PeerJ **22**(6), e5982 (2018)
41. Lu, C.-F., et al.: Machine learning-based radiomics for molecular subtyping of gliomas. Clin. Cancer Res. **24**(18), 4429–4436 (2018)
42. Tixier, F., et al.: Preoperative MRI-radiomics features improve prediction of survival in glioblastoma patients over MGMT methylation status alone. Oncotarget **10**(6), 660–672 (2019)
43. Bae, S., et al.: Radiomic MRI phenotyping of glioblastoma: improving survival prediction. Radiology **289**(3), 797–806 (2018)
44. Wen, P.Y., et al.: Updated response assessment criteria for high- grade gliomas: response assessment in neuro-oncology working group. JCO **28**(11), 1963–1972 (2010)
45. Hsieh, K.L.-C., Chen, C.-Y., Lo, C.-M.: Radiomic model for predicting mutations in the isocitrate dehydrogenase gene in glioblastomas. Oncotarget **8**(28), 45888–45897 (2017)
46. Liao, X., Cai, B., Tian, B., Luo, Y., Song, W., Li, Y.: Machine-learning based radiogenomics analysis of MRI features and metagenes in glioblastoma multiforme patients with different survival time. J. Cell. Mol. Med. **23**(6), 4375–4385 (2019)
47. Xi, Y.-B., et al.: Radiomics signature: a potential biomarker for the prediction of MGMT promoter methylation in glioblastoma. J. Magn. Reson. Imaging **47**(5), 1380–1387 (2018)
48. Gutman, D.A., et al.: Somatic mutations associated with MRI-derived volumetric features in glioblastoma. Neuroradiology **57**(12), 1227–1237 (2015)
49. Gevaert, O., et al.: Glioblastoma multiforme: exploratory radiogenomic analysis by using quantitative image features. Radiology **273**(1), 168–174 (2014)
50. Hu, L.S., et al.: Radiogenomics to characterize regional genetic heterogeneity in glioblastoma. Neuro Oncol. **19**(1), 128–137 (2017)

A Survey on Recent Advancements for AI Enabled Radiomics in Neuro-Oncology

Syed Muhammad Anwar[1,3]([✉]), Tooba Altaf[2], Khola Rafique[3], Harish RaviPrakash[1], Hassan Mohy-ud-Din[4], and Ulas Bagci[1]

[1] Center for Research in Computer Vision, University of Central Florida, Orlando, FL 32816, USA
s.anwar@knights.ucf.edu
[2] Department of Computer Science, University of Wah, Wah, Pakistan
[3] Department of Software Engineering, University of Engineering and Technology, Taxila 47050, Pakistan
[4] Syed Babar Ali School of Science and Engineering, Lahore University of Management Sciences (LUMS), Lahore, Pakistan

Abstract. Artificial intelligence (AI) enabled radiomics has evolved immensely especially in the field of oncology. Radiomics provide assistance in diagnosis of cancer, planning of treatment strategy, and prediction of survival. Radiomics in neuro-oncology has progressed significantly in the recent past. Deep learning has outperformed conventional machine learning methods in most image-based applications. Convolutional neural networks (CNNs) have seen some popularity in radiomics, since they do not require hand-crafted features and can automatically extract features during the learning process. In this regard, it is observed that CNN based radiomics could provide state-of-the-art results in neuro-oncology, similar to the recent success of such methods in a wide spectrum of medical image analysis applications. Herein we present a review of the most recent best practices and establish the future trends for AI enabled radiomics in neuro-oncology.

Keywords: Radiomics · Neuro-oncology · Classification · Deep learning

1 Introduction

Brain and other central nervous system (CNS) tumors account for the second most common cancer affecting children, and the third most common cancer affecting adolescents and young adults [1,2]. There are approximately 700,000 people with primary brain or CNS tumors in the United States alone [1]. Treatment is dependent on multiple factors including age, gender, tumor size and location, etc. The standard approach in most cases is to surgically remove the tumor via craniotomy [3]. However, some tumors cannot be surgically removed

H. Mohy-ud-Din and S. Rathore (Eds.): RNO-AI 2019, LNCS 11991, pp. 24–35, 2020.
https://doi.org/10.1007/978-3-030-40124-5_3

and the treatment then relies on radiation therapy. Rigorous planning is necessary to determine the exact tumor volume and a buffer region surrounding the tumor which has to be treated to prevent growth from left over malignant cells. The accurate planning of resection and radiation area is challenging owing to the difficulty in determining the exact tumor dimensions. For manual segmentation (delineation), the radiologists need to carefully analyze a large amount of radiology images. To ease the load on radiologists, computational methods to automatically extract quantitative features (aka radiomics) from radiological scans have been proposed.

Radiomics comprises of numerous significant disciplines, including radiology, computer vision, and machine learning. The objective is the recognition of quantitative imaging features with an anticipation of significant clinical results in prognosis and analysis of certain treatment strategies [61]. The information provided by radiology scans is processed with the help of quantitative image analysis (QIA) for identifying patterns in radiology scans in a way that human eye may not achieve. Different steps includes, acquire and store images, segment and identify region of interest (ROIs), extract features, build and validate model, and integration of these process into clinical decision support system. The resultant units of data from QIA may be called quantitative imaging bio-markers depending on their predictive powers. A huge amount of information is captured during clinical imaging but the underlying data, in most cases, have been reported in subjective and qualitative terms. Specifically, radiomics in neuro-oncology aims to revamp the brain tumor treatment paradigm by extracting quantitative features from brain scans (MRI). Data is mined via multiple machine learning algorithms and can potentially be used as imaging bio-markers to distinguish intra-tumoral dynamics during treatment [20]. With the increase in number of reported cancer cases, analytic methods for imaging have revealed new understandings about initial treatment response, risk factors, and optimal treatment approaches [30,60]. Image-based models are turning into a significant empowering innovation that allow investigation and approval of selected quantitative features.

The recent advancements, particularly in Artificial Intelligence (AI), are impacting major technological and scientific fields. To keep up with these advancements, medical science is adapting new methodologies for improving diagnosis and treatment of various clinical conditions [32]. In clinical setting, imaging has played a vital role for a long time by helping physicians in diagnostic and treatment related decisions making [5]. However, over a passage of time, medical imaging has evolved from just being a diagnostic tool and is now beginning to take a critical role in precision medicine for tasks such as screening, diagnosis, guided treatment, and assessing the disease recurrence likelihood [18]. The emerging field of radiomics in oncology has helped in developing a latent solution for tumor characterization by extracting a large number of features from medical images [31,44]. Attributes that can be used in assessment of tissue appearance by radiologist are of great importance and can be used in the development of medical imaging analysis techniques. Some common examples of such attributes include texture, intensity, and morphology. *Texture* can be defined as the spatial variation of pixel intensities within an image, and is known to be particularly sensitive for the assessment of pathology [19]. Visual assessment of texture is however, particularly subjective. Additionally, it is known that human observers possess limited sensitivity to textural patterns, whereas computational texture analysis techniques can be significantly more sensitive to such changes [16]. For image classification, numerous computer vision algorithms depend on extracting native characteristics form images. These features are handcrafted with an eye for resolving explicit issues like obstructions and variations in scale and brightness. The design of handcrafted features often involves finding the right trade-off between accuracy and computational efficiency [40]. In contrast, deep learning (DL) methods have a huge potential to replace conventional machine learning methods for automatically extracting imaging features which are more efficient and give state-of-the-art performance in a large number of applications already. In the following, we present a review of methods relying on handcrafted features and those using DL and analyze the future direction for AI enabled radiomics in neuro-oncology.

2   Radiomics Using Handcrafted Features

A general pipeline for radiomics in neuro-oncology is shown in Fig. 1. Different radiomics features are extracted from medical images and then machine learning classifiers are used to detect diseases such as brain tumor. These radiomics features are either extracted in a hand crafted manner or through DL. The top layer (Fig. 1) shows how handcrafted features are used with different radiology image inputs. The feature extraction stage (also known as conventional radiomics approach) relies on selecting features from various domains such as texture, intensity/density, and frequency (e.g., wavelet). Different machine learning classifiers (support vector machines (SVMs) and logistic regression (LR)) are used for analysis of these features and results are analyzed using performance parameters (such as accuracy and receiver operating characteristics (ROC)). Whereas, in deep learning the model chooses the appropriate features, allowing feature learning, which can then be used directly for classification/regression. These learned features can also be used with other classifiers such as SVM.

One of the approaches employed to extract radiomic features is called local binary patterns (LBPs) [4,17,21,45], where binary word encoding is used to incorporate relationship between pixels and their neighbours. This enables LBP to detect patterns in the image irrespective of contrast variations. LBP feature extractor is known for its efficiency in utilizing the computation power, but its effectiveness reduces with an increase in noise in the image [43]. Another commonly used method to extract radiomics features is Histogram of Oriented Gradient (HOG) [50,57] where the number of oriented gradient occurrences in certain image regions are counted to create a histogram. Depending on the application, different regions can be used to capture local shape and edge information from the images, which is further converted into a feature vector using the HOG

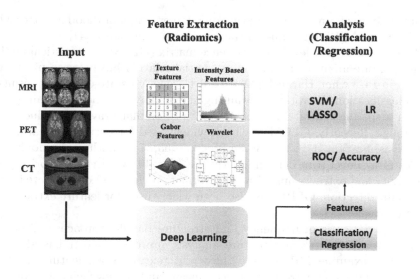

Fig. 1. A pipeline of steps for radiomics in radiology using handcrafted and DL based features

Table 1. Radiomics in neuro-oncology using handcrafted features.

Method (year)	Features	Classifier	Accuracy/specificity /sensitivity
[45] (2019)	LBP	SVM	97.02/94.28/98.48
[10] (2019)	Discrete wavelet transform/bag of words	SVM	100/-/-
[21] (2019)	Fusion of LBP, GLCM and GLRL	Ensemble	97.14/-/-
[50] (2019)	GLCM + PHOG + intensity-based + modified CLBP	PSO-SVM	98.36/97.83/99.17
[46] (2019)	GLCM + GLRL + gray level size zone matrix + first order statistics + shape descriptors	SVM + LASSO	90/-/-
[57] (2018)	GLCM + GLRL + HOG + neighbourhood grey-tone difference matrix	RUSBoost ensemble classifier	73.2/-/-
[17] (2018)	LBP	SVM	95/94/96
[59] (2018)	GLCM + GLRLM features + gabor descriptor	SVM	71.02/-/-
[36] (2018)	Multiple hand crafted	Radiomics nomogram + ROC	81.52/-/-
[26] (2018)	Radiomics signature (Lasso-Cox regression)	Thresholding	95/-/-
[37] (2018)	Statistical + histogram features + GLCM + GLRLM + GLZLM	Logistic regression	89/96/85
[48] (2018)	GLCM + GLRL + fractal dimensions + wavelet filtered GLCM	Logistic regression	95/-/-
[34] (2018)	Statistical + shape-based + texture + wavelet	LASSO Cox regression model	82.3/-/-
[19] (2017)	Gabor texture descriptor	SVM	97.5-92-99
[4] (2017)	LBP + HOG	Random forest	83.77/-/-

descriptor. It was found that operating on a larger neighbourhood is better when using HOG for MR images due to the low intensity variance [41].

The first use of gray level co-occurrence matrix (GLCM), a statistical method used for texture analysis by examining spatial relationship between pixels, was recorded in 1973 when Haralick [22] used it to generate state-of-the-art results in image classification. It works by counting the number of times a certain pair of pixels in a specific spatial relationship and with similar gray scale values occur in an image. Recently, GLCM has been widely used for extracting features for disease classification [11,13,21,34,36,37,48,50,55,57]. Another commonly used method, Gray Level Run Length Matrix (GLRL) [49], works on the principle of connectivity and extracts quantitative information (lengths) of connected pixels in a specific direction. GLRL has also been widely used for feature extraction in radiomics studies [59].

As a special class of frequency and structure based approaches, Gabor filter has shown to be popular texture analysis approach, and it has also been employed to examine MR scans to filter out texture-based features such as, smoothness, kurtosis, entropy, contrast, mean, and homogeneity. Gabor filter works especially well for images with uniform patterns. Medical images usually possess pixels with low variance of intensity levels and uniform orientation. Hence, Gabor filter may outperform other texture-based descriptors in case it has the capability to encode narrow bands of occurrences and orientations. Gabor filter is also good at examining structure differentiation that are caused by cancerous cells in MR images making it ideal for medical imaging data. For these reasons, these filters have been used for extracting radiomic features in multiple studies [19,39,59]. Radiomics is also applied successfully in other diagnostic applications, some recent works are summarized in Table 1 highlighting the features and classifiers used. After extracting radiomics features by using various descriptors, a classifier assigns a particular class to the patient image. Most methods (Table 1) use support vector machine as a classifier. Other methods include least absolute shrinkage and selection operator (LASSO), random forest and logistic regression. It is important to observe here that there is a wide array of descriptors available and hence requires a lot of handcrafting to chose the most appropriate features. An automated system that can learn features from raw input data could help in providing more generalized results for the increasing number of radiology studies.

3   Radiomics Using Deep Learning

Recently, the most widely used machine learning techniques are based on deep learning, where various functions are used to transform the data into a hierarchical representation [47]. DL has gained wide attention in image categorization, image recognition, speech recognition and natural language processing, and medical image analysis [8,56]. One major advantage of DL is the fact that features are extracted directly from raw data allowing feature learning [27]. DL is also found successful in solving complex problems with limited data, using transfer learning wherein a model trained on one type of data is used to train a different complex task [53]. On the flip side, DL is generally known to be more successful in solving problems where large data sets are available [27], although methods that work for limited data are emerging [54].

There are two popular approaches (Fig. 1) used in DL - training a network and extracting the features to use with a simple machine learning classifier and training an end-to-end network that incorporates the classification/regression task in it's learning. An example of the former is the work by Nie et al. [42]. The authors proposed a multi-channel artificial intelligence enabled radiomics for predicting patients survival time in neuro-oncological applications. First, the proposed technique used three-dimensional convolutional neural networks for extracting high level features from multi-modal MR images. In the second step, those features along with patients personal details and medical history were fed to an SVM for predicting the survival time. The proposed method achieved state-of-the-art results with an accuracy of 90.66%. Chang et al. [14] proposed an end-to-end trained residual convolutional network to diagnose isocitrate dehydrogenase (IDH) mutations in people suffering from grade II-IV gliomas. The diagnosis of IDH mutations could assist radiologists in the treatment of patients suffering from gliomas. The network was trained on multi-institutional clinical MRI data and different techniques like random rotation and zooming was used to reduce data over-fitting. The proposed network gave an accuracy and area under the curve (AUC) of 82.8% and 0.90 respectively on training data, 83% and 0.93 on validation data, and 85.7% and 0.94 on testing data respectively. This artificial intelligence based radiomics is currently considered as the largest study for the prediction of IDH mutations. In [33] proposed deep learning enabled radiomics for survival prediction of patients suffering from glioblastoma multiforme. The proposed technique used transfer learning for predicting patients survival. The features were extracted from MR images using conventional and deep learning methods. The features extracted from deep learning were fed to LASSO Cox model for predicting patients survival. The proposed technique also required demographic information such as age and Karnofsky Performance Score. The technique has some limitations as it was designed for small dataset and, also the relation between features and patients genetic details were not investigated. The results showed that deep learning based radiomics achieved better prognosis than conventional machine learning based radiomics.

There are various methods reported in literature related to brain diseases that are based on both conventional features and DL based methods [7,9,15,24,25]. In [58], authors combines fully convolutional neural network with conditional random field (CRF). The technique used image patches for training fully convolutional neural network and 2D image slices: coronal, sagital and axial, for training CRF as recurrent neural network. Then image slices were used to fine tune both networks. The experiments were carried out on BraTS 2013, 2015 and 2016 data sets [12,38]. This study trained three segmentation models using both image patches and slices, and it has been observed that slice by slice segmentation was computationally more effective than segmentation using image patches. This method worked well for 2D images but did not perform well for 3D volumes. Cascaded anisotropic convolutional neural networks were employed to segment multi-class brain tumor [52]. The developed technique treated all three classes (core, enhancing, and whole) separately, and three different network architectures were designed and concatenated. Anisotropic network was designed to resolve model complexity arising from the use of large receptive fields. Residual connections were employed for robust training and segmentation performance. The model was tested on BraTS 2017 dataset [12] and achieved the dice scores (DSC) of 0.7831, 0.8739, and 0.7748 for enhancing, whole, and core tumor regions respectively. The experiments showed that this setup has made training easier and reduced false positives. But this technique is not end-to-end and consumes more time in training and testing than other techniques.

Table 2. DL based radiomics approaches in neuro-oncology.

Method (year)	Dataset	Architecture	Task	Performance parameter
Nie et al. (2019)	Clinical (glioma) images	CNN + SVM	Survival prediction	Accuracy: 90.66%
Chang et al. (2018)	Clinical MR images	Res-Net	IDH phenotype prediction	Accuracy: 89.1% (AUC = 0.95)
Zhao et al. (2018)	BraTS	CNN+ CRF-RNN	Tumor segmentation	DSC: Whole-0.82, Core-0.72, Enhanced-0.62
Wang et al. (2017)	BraTS 2017	Cascaded CNN	Tumor segmentation	DSC: Whole-0.87, Core-0.77, Enhanced-0.78
Alex et al. (2017)	BraTS 2017	CNN + texture features	Tumor segmentation	DSC: Whole-0.83, Core-0.69, Enhanced-0.72
Havaei et al. (2017)	BraTS 2013	Cascaded CNN	Tumor segmentation	DSC: Whole-0.81, Core-0.72, Enhance-0.58
Lao et al. (2017)	Clinical data	CNN + LASSO Cox	Overall survival	C-index = 0.739
Liu et al. (2017)	BraTS 2015	CNN	Tumor segmentation	DSC: Core-0.75, Enhanced-0.81

A 23-layered fully convolutional neural network was proposed for segmentation of gliomas from MRI [6]. Texture analysis including first order texture features and shape-based features was used for the prediction of patients survival. The designed algorithm was trained on 2D slices extracted from patients MRI volume. The proposed network gave the survival prediction accuracy of 52% and 47% on BraTS 2017 validation and testing dataset respectively. The achieved DSC on BraTS 2017 for whole tumor, tumor core and enhanced region was 0.83, 0.69 and 0.72, respectively. A novel CNN architecture was proposed which incorporated both dual pathway and cascaded architecture for radiomics in neuro-oncology [23]. The output of cascaded architecture was fed to dual pathway network improved prediction accuracy. The convolutional neural network predicts labels independent of its neighboring pixels which limits its capability for producing accurate results. The cascaded architecture output made it possible for the proposed CNN to incorporate the influence of neighboring pixels. This variation of convolutional neural network increased the speed by 40 folds and incorporated both local and global features. The fully connected layer of the proposed network architecture was designed in convolutional manner. Two phase training technique was used for accurate delineation of brain tumor and it was tested on BraTS 2013 dataset. The proposed architecture worked well for two-dimensional data but slows down in case of three-dimensional data.

An algorithm was devised using convolutional neural network for segmentation of brain metastases from MRI [35]. Image patches were fed to the network for voxel-wise classification which made the setup efficient for segmenting small lesions. Although the network was designed for mono-modality imaging, nonetheless it was also tested on multi-modality dataset (BraTS), where DSC values of 0.75 and 0.81 were achieved on core and enhanced tumors, respectively. The network was trained on pre-defined parameters which made it more robust. The performance of this network architecture could be improved by readjusting patch size and hyper-parameters, however. This AI-enabled radiomics in neuro-oncology could help in treatment strategy planning for brain metastases. In [29], authors proposed *deepmedic* platform, a dual pathway network incorporating local and global features, for segmenting brain tumors. Conditional random fields were used as a post-processing step to reduce the number of false positives. An improvement to deepmedic was proposed using residual connections, and performance was evaluated on a small dataset (BraTS 2015) to make this approach more flexible [28]. This simplified approach achieved good results on BraTS 2016 as well, where DSC score by using 75% of the data was 91.4, 83.1 and 79.4 for whole tumor, core, and enhanced tumor regions. Table 2 gives a summary of the methods in segmentation and prediction. Although the results of DL are promising, the methodology suffers with the black box paradigm. The feature learning process is still not transparent and the aim of achieving generalization is still to be achieved.

4   Discussion and Conclusion

AI-enabled radiomics is making significant progress in neuro-oncology and similar applications, with performance better than conventional approaches. It aids radiologists in making an accurate prognosis leading to better treatment strategy. An important consideration is finding the right hand-crafted features, as the results have shown that these features can significantly effect the overall outcome of the method. A possible solution to this impediment is to use DL which is known to learn the right features in an automated fashion, when a reasonable amount of training data is present. It is observed that DL based methods are able to produce state-of-the-art results. Both radiomics and DL fields are currently developing at a very fast pace. It is believed that they will work together in future resulting in the development of AI enabled radiomics that will transform not only prognosis and diagnosis, but also how treatment planning and analysis of disease recurrence works in oncology.

Various tumor types may appear similar on radiology images, but the molecular characteristics of different malignant parts vary. Moreover tumor phenotype changes with the passage of time, hence biopsies cannot provide much information. Hence, personalized medicine predicts different results and more effective treatments, in the light of improved serum, tissue, and imaging bio-markers [51]. Radiomics can assist by evaluating the imaging bio-markers that would identify the tumor signature clearly and hence show the tumor function and evolution. These statistics will help multi-disciplinary oncology members to develop a highly personalized curative plan for individuals based on the information of exactly how that specific patients cancer is expected to behave. Interpretable DL will help in identifying the right radiomic features improving upon the hand-crafted features based methods. For precision and accuracy in this challenging area, more interpretation and explainability is required for the underlying DL-based models.

References

1. American Brain Tumor Association. http://abta.pub30.convio.net/about-us/news/brain-tumor-statistics/. Accessed 07 Jan 2019
2. Cancer.net.   https://www.cancer.net/cancer-types/brain-tumor/statistics. Accessed 07 Jan 2019
3. UCSF health. https://www.ucsfhealth.org/conditions/brain_tumor/treatment. html. Accessed 07 Jan 2019
4. Abbasi, S., Tajeripour, F.: Detection of brain tumor in 3D MRI images using local binary patterns and histogram orientation gradient. Neurocomputing **219**, 526–535 (2017)
5. Aerts, H.J., et al.: Decoding tumour phenotype by noninvasive imaging using a quantitative radiomics approach. Nat. Commun. **5**, 4006 (2014)
6. Alex, V., Safwan, M., Krishnamurthi, G.: Automatic segmentation and overall survival prediction in gliomas using fully convolutional neural network and texture analysis. In: Crimi, A., Bakas, S., Kuijf, H., Menze, B., Reyes, M. (eds.) BrainLes 2017. LNCS, vol. 10670, pp. 216–225. Springer, Cham (2018). https://doi.org/10.1007/978-3-319-75238-9_19
7. Altaf, T., Anwar, S.M., Gul, N., Majeed, M.N., Majid, M.: Multi-class alzheimer's disease classification using image and clinical features. Biomed. Signal Process. Control **43**, 64–74 (2018)
8. Anwar, S.M., Majid, M., Qayyum, A., Awais, M., Alnowami, M., Khan, M.K.: Medical image analysis using convolutional neural networks: a review. J. Med. Syst. **42**(11), 226 (2018)
9. Ateeq, T., et al.: Ensemble-classifiers-assisted detection of cerebral microbleeds in brain MRI. Comput. Electr. Eng. **69**, 768–781 (2018)
10. Ayadi, W., Elhamzi, W., Charfi, I., Atri, M.: A hybrid feature extraction approach for brain MRI classification based on bag-of-words. Biomed. Signal Process. Control **48**, 144–152 (2019)

11. Bagci, U., Yao, J., Miller-Jaster, K., Chen, X., Mollura, D.J.: Predicting future morphological changes of lesions from radiotracer uptake in 18F-FDG-PET images. PLoS ONE **8**(2), e57105 (2013)
12. Bakas, S., et al.: Advancing the cancer genome atlas glioma MRI collections with expert segmentation labels and radiomic features. Sci. Data **4**, 170117 (2017)
13. Buty, M., Xu, Z., Gao, M., Bagci, U., Wu, A., Mollura, D.J.: Characterization of lung nodule malignancy using hybrid shape and appearance features. In: Ourselin, S., Joskowicz, L., Sabuncu, M.R., Unal, G., Wells, W. (eds.) MICCAI 2016. LNCS, vol. 9900, pp. 662–670. Springer, Cham (2016). https://doi.org/10.1007/978-3-319-46720-7_77
14. Chang, K., et al.: Residual convolutional neural network for the determination of IDH status in low-and high-grade gliomas from MR imaging. Clin. Cancer Res. **24**(5), 1073–1081 (2018)
15. Farooq, A., Anwar, S., Awais, M., Alnowami, M.: Artificial intelligence based smart diagnosis of alzheimer's disease and mild cognitive impairment. In: 2017 International Smart cities conference (ISC2), pp. 1–4. IEEE (2017)
16. Fetit, A.E., et al.: Radiomics in paediatric neuro-oncology: a multicentre study on MRI texture analysis. NMR Biomed. **31**(1), e3781 (2018)
17. Giacalone, M., et al.: Local spatio-temporal encoding of raw perfusion MRI for the prediction of final lesion in stroke. Med. Image Anal. **50**, 117–126 (2018)
18. Giardino, A., et al.: Role of imaging in the era of precision medicine. Acad. Radiol. **24**(5), 639–649 (2017)
19. Gilanie, G., Bajwa, U.I., Waraich, M.M., Habib, Z., Ullah, H., Nasir, M.: Classification of normal and abnormal brain MRI slices using gabor texture and support vector machines. Signal Image Video Process. **12**(3), 479–487 (2018)
20. Gillies, R.J., Kinahan, P.E., Hricak, H.: Radiomics: images are more than pictures, they are data. Radiology **278**(2), 563–577 (2015)
21. Gupta, N., Bhatele, P., Khanna, P.: Glioma detection on brain MRIs using texture and morphological features with ensemble learning. Biomed. Signal Process. Control **47**, 115–125 (2019)
22. Haralick, R.M., Shanmugam, K., et al.: Textural features for image classification. IEEE Trans. Syst. Man Cybern. **6**, 610–621 (1973)
23. Havaei, M., et al.: Brain tumor segmentation with deep neural networks. Med. Image Anal. **35**, 18–31 (2017)
24. Hussain, S., Anwar, S.M., Majid, M.: Brain tumor segmentation using cascaded deep convolutional neural network. In: 2017 39th Annual International Conference of the IEEE Engineering in Medicine and Biology Society (EMBC), pp. 1998–2001. IEEE (2017)
25. Hussain, S., Anwar, S.M., Majid, M.: Segmentation of glioma tumors in brain using deep convolutional neural network. Neurocomputing **282**, 248–261 (2018)
26. Jiang, Y., et al.: Radiomics signature of computed tomography imaging for prediction of survival and chemotherapeutic benefits in gastric cancer. EBioMedicine **36**, 171–182 (2018)
27. Kamilaris, A., Prenafeta-Boldú, F.X.: Deep learning in agriculture: a survey. Comput. Electron. Agric. **147**, 70–90 (2018)
28. Kamnitsas, K., et al.: DeepMedic for brain tumor segmentation. In: Crimi, A., Menze, B., Maier, O., Reyes, M., Winzeck, S., Handels, H. (eds.) BrainLes 2016. LNCS, vol. 10154. Springer, Cham (2016). https://doi.org/10.1007/978-3-319-55524-9_14
29. Kamnitsas, K., et al.: Efficient multi-scale 3D CNN with fully connected CRF for accurate brain lesion segmentation. Med. Image Anal. **36**, 61–78 (2017)

30. Kotrotsou, A., Zinn, P.O., Colen, R.R.: Radiomics in brain tumors: an emerging technique for characterization of tumor environment. Magn. Reson. Imaging Clin. **24**(4), 719–729 (2016)
31. Kumar, V., et al.: Radiomics: the process and the challenges. Magn. Reson. Imaging **30**(9), 1234–1248 (2012)
32. Lambin, P., et al.: Radiomics: the bridge between medical imaging and personalized medicine. Nat. Rev. Clin. Oncol. **14**(12), 749 (2017)
33. Lao, J., et al.: A deep learning-based radiomics model for prediction of survival in glioblastoma multiforme. Sci. Rep. **7**(1), 10353 (2017)
34. Liu, X., et al.: A radiomic signature as a non-invasive predictor of progression-free survival in patients with lower-grade gliomas. NeuroImage: Clin. **20**, 1070–1077 (2018)
35. Liu, Y., et al.: A deep convolutional neural network-based automatic delineation strategy for multiple brain metastases stereotactic radiosurgery. PLoS ONE **12**(10), e0185844 (2017)
36. Liu, Z., et al.: Radiomics analysis allows for precise prediction of epilepsy in patients with low-grade gliomas. NeuroImage: Clin. **19**, 271–278 (2018)
37. Lohmann, P., et al.: Combined FET PET/MRI radiomics differentiates radiation injury from recurrent brain metastasis. NeuroImage: Clin. **20**, 537–542 (2018)
38. Menze, B.H., et al.: The multimodal brain tumor image segmentation benchmark (BRATS). IEEE Trans. Med. Imaging **34**(10), 1993–2024 (2014)
39. Nabizadeh, N., Kubat, M.: Brain tumors detection and segmentation in MR images: gabor wavelet vs. statistical features. Comput. Electr. Eng. **45**, 286–301 (2015)
40. Nanni, L., Ghidoni, S., Brahnam, S.: Handcrafted vs. non-handcrafted features for computer vision classification. Pattern Recogn. **71**, 158–172 (2017)
41. Nanni, L., Salvatore, C., Cerasa, A., Castiglioni, I., Initiative, A.D.N., et al.: Combining multiple approaches for the early diagnosis of alzheimer's disease. Pattern Recogn. Lett. **84**, 259–266 (2016)
42. Nie, D., et al.: Multi-channel 3D deep feature learning for survival time prediction of brain tumor patients using multi-modal neuroimages. Sci. Rep. **9**(1), 1103 (2019)
43. Ojala, T., Pietikäinen, M., Mäenpää, T.: Multiresolution gray-scale and rotation invariant texture classification with local binary patterns. IEEE Trans. Pattern Anal. Mach. Intell. **7**, 971–987 (2002)
44. Parmar, C., et al.: Robust radiomics feature quantification using semiautomatic volumetric segmentation. PLoS ONE **9**(7), e102107 (2014)
45. Polepaka, S., Rao, C.S., Mohan, M.C.: IDSS-based two stage classification of brain tumor using SVM. Health Technol., 1–10 (2019)
46. Qian, Z., et al.: Differentiation of glioblastoma from solitary brain metastases using radiomic machine-learning classifiers. Cancer Lett. **451**, 128–135 (2019)
47. Schmidhuber, J.: Deep learning in neural networks: an overview. Neural Netw. **61**, 85–117 (2015)
48. Shen, C., et al.: Building CT radiomics based nomogram for preoperative esophageal cancer patients lymph node metastasis prediction. Transl. Oncol. **11**(3), 815–824 (2018)
49. Singh, K.H.R.: A comparison of gray-level run length matrix and gray-level co-occurrence matrix towards cereal grain classification. Int. J. Comput. Eng. Technol. (IJCET) **7**(6), 9–17 (2016)
50. Song, G., et al.: A noninvasive system for the automatic detection of gliomas based on hybrid features and PSO-KSVM. IEEE Access **7**, 13842–13855 (2019)

51. Subramanyam, M., Goyal, J.: Translational biomarkers: from discovery and development to clinical practice. Drug Discov. Today: Technol. **21**, 3–10 (2016)
52. Wang, G., Li, W., Ourselin, S., Vercauteren, T.: Automatic brain tumor segmentation using cascaded anisotropic convolutional neural networks. In: Crimi, A., Bakas, S., Kuijf, H., Menze, B., Reyes, M. (eds.) BrainLes 2017. LNCS, vol. 10670, pp. 178–190. Springer, Cham (2018). https://doi.org/10.1007/978-3-319-75238-9_16
53. Weiss, K., Khoshgoftaar, T.M., Wang, D.: A survey of transfer learning. J. Big data **3**(1), 9 (2016)
54. Wong, K.C., Syeda-Mahmood, T., Moradi, M.: Building medical image classifiers with very limited data using segmentation networks. Med. Image Anal. **49**, 105–116 (2018)
55. Wu, S., et al.: Development and validation of an MRI-based radiomics signature for the preoperative prediction of lymph node metastasis in bladder cancer. EBioMedicine **34**, 76–84 (2018)
56. Yasaka, K., Akai, H., Kunimatsu, A., Kiryu, S., Abe, O.: Deep learning with convolutional neural network in radiology. Japan. J. Radiol. **36**(4), 257–272 (2018)
57. Zhang, Z., et al.: A predictive model for distinguishing radiation necrosis from tumour progression after gamma knife radiosurgery based on radiomic features from MR images. Eur. Radiol. **28**(6), 2255–2263 (2018)
58. Zhao, X., Wu, Y., Song, G., Li, Z., Zhang, Y., Fan, Y.: A deep learning model integrating FCNNS and CRFS for brain tumor segmentation. Med. Image Anal. **43**, 98–111 (2018)
59. Zhou, H., et al.: Diagnosis of distant metastasis of lung cancer: based on clinical and radiomic features. Transl. Oncol. **11**(1), 31–36 (2018)
60. Zhou, M., Chaudhury, B., Hall, L.O., Goldgof, D.B., Gillies, R.J., Gatenby, R.A.: Identifying spatial imaging biomarkers of glioblastoma multiforme for survival group prediction. J. Magn. Reson. Imaging **46**(1), 115–123 (2017)
61. Zhou, M., et al.: Radiomics in brain tumor: image assessment, quantitative feature descriptors, and machine-learning approaches. Am. J. Neuroradiol. **39**(2), 208–216 (2018)

Deep Radiomic Features from MRI Scans Predict Survival Outcome of Recurrent Glioblastoma

Ahmad Chaddad[1,4](✉), Mingli Zhang[2], Christian Desrosiers[3],
and Tamim Niazi[1]

[1] Department of Oncology, McGill University, Montreal, Canada
ahmad.chaddad@mail.mcgill.ca
[2] Montreal Neurological Institute, McGill University, Montreal, Canada
[3] The Laboratory for Imagery, Vision and Artificial Intelligence,
ETS, Montreal, Canada
[4] School of Artificial Intelligence, Guilin University of Electronic Technology,
Guilin, China

Abstract. This paper proposes to use deep radiomic features (DRFs) from a convolutional neural network (CNN) to model fine-grained texture signatures in the radiomic analysis of recurrent glioblastoma (rGBM). We use DRFs to predict survival of rGBM patients with preoperative T1-weighted post-contrast MR images (n = 100). DRFs are extracted from regions of interest labelled by a radiation oncologist and used to compare between short-term and long-term survival patient groups. Random forest (RF) classification is employed to predict survival outcome (i.e., short or long survival), as well as to identify highly group-informative descriptors. Classification using DRFs results in an area under the ROC curve (AUC) of 89.15% (p < 0.01) in predicting rGBM patient survival, compared to 78.07% (p < 0.01) when using standard radiomic features (SRF). These results indicate the potential of DRFs as a prognostic marker for patients with rGBM.

Keywords: Classification · Deep learning · Radiomics · rGBM

1 Introduction

Gliomas are the most common type of primary brain tumor in adults. They can be classified by histolopathological features into four grades (I, II, III or IV) as mentioned in the World Health Organization (WHO). Grade I glioma correspond to non-invasive tumors, grade II/III to low/intermediate-grade gliomas, and grade IV to aggressive malignant tumors called glioblastoma (GBM) [1]. GBM is a devastating disease of the primary central nervous system with ubiquitously poor outcome and a median survival of less than 15 months [2]. Most patients relapse within months, after which there are limited options for further treatment [3]. Improvement of patient survival represents one of the biggest challenges for recurrent GBM (rGBM).

Radiomics analysis for the automated prognosis in brain tumor patients uses a wide range of imaging features computed from region of interest (ROI) as input to a

© Springer Nature Switzerland AG 2020
H. Mohy-ud-Din and S. Rathore (Eds.): RNO-AI 2019, LNCS 11991, pp. 36–43, 2020.
https://doi.org/10.1007/978-3-030-40124-5_4

classifier model [4, 5]. Standard radiomics approaches rely on a variety of hand-crafted features, for instance, based on histograms of intensity, grey level co-occurrence matrix (GLCM), neighborhood gray-tone difference matrix (NGTDM) and gray-level zone size matrix (GLZSM). In the last years, CNNs have achieved state-of-art performance for a wide range of image classification tasks [6]. A CNN is a multi-layered architecture that incorporates spatial context and weight sharing between pixels or voxels. Unlike standard radiomic techniques, which rely on hand-crafted features to encode images, CNNs basically learn image representations that are convenient for classification tasks, directly from training data [7]. The main components of CNN are stacks of different types of specialized layers (i.e., convolutional, activation, pooling, fully connected, softmax, etc.) that are interconnected and whose weights are trained using the back-propagation algorithm with some tuning functions (e.g., stochastic gradient descent with momentum). A common limitation of CNNs, when employed directly for pre-diction, is their requirement for large training sets which are often unavailable. An alternative strategy uses CNNs as a general technique to extract a reduced set of informative image features that are then fed to a standard classifier model. Since the CNN feature extractor and classifier are learned using separate training sets, this strategy is less prone to overfitting when data is limited. Recently, an approach using deep CNN features with a support vector machine (SVM) classifier was shown useful for predicting the survival of limited GBM patients [8]. Despite this success, the exploitation of multiscale features across different CNN layers as learnable texture descriptors remains limited. In this work, we argue that tumor progression can effec-tively be captured by texture descriptors learned using a CNN. Hence, we propose to extract deep radiomics features (DRFs) from a 3D-CNN with 41 texture quantifier functions, and use these features as input to a random forest (RF) model for predicting the survival of rGBM patients.

In our previous work [9, 10], a similar strategy was proposed for differentiating between normal brain aging and Alzheimer's disease (AD) in MRI data. Specifically, we used the entropy of convolutional feature maps as texture descriptors for classifying normal control versus AD subjects. In contrast, the current work considers a broader set of 41 quantifier functions to compute texture descriptors for predicting rGBM patient survival. We hypothesize that texture within CNN layers captures important charac-teristics of tumor heterogeneity which are highly-relevant for predicting clinical out-come (i.e. survival). Additionally, we address the problem of limited training data using a transfer learning strategy where the 3D-CNN to extract features is pretrained on MRI images for other prediction tasks.

2 Materials and Methods

We describe our deep feature model DRF based on deep 3D CNNs and 41 standard radiomic features (SRFs), as shown in Fig. 1. Post contrast T1-weighted images are first acquired for rGBM patients. Gross-total-resection (i.e., tumor ROI) are manually labelled in each scan using 3D Slicer tool. A set of 41 texture descriptors is then extracted from labelled images in two different ways: (a) applying standard quantifier functions on the multiscale feature maps of a pretrained 3D CNN; (b) applying these

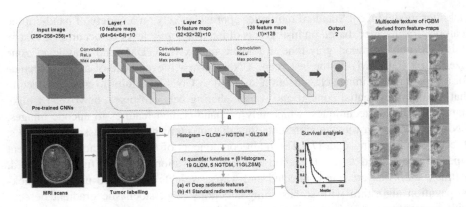

Fig. 1. Workflow of the proposed method to predict survival of rGBM patients. (1) identification and labelling GBM tumors in post contrast T1-WI MR images; (2) 41 quantifier functions encode the histogram, GLCM, NGTDM and GLZSM of feature maps (e.g. multiscale texture of rGBM) in layer 2 and 3 that derived from pretrained 3D CNN. These 41 DRF features in (a) compared to the 41 standard radiomic features using the log-rank, Kaplan-Meier estimator and RF classifier.

quantifier functions directly on the original ROIs. Various analyses are considered to evaluate the usefulness of DRFs to predict survival outcome. To identify features which are differentially enriched in short or long survivors, we separate patients based on the median value of radiomic features and assess survival difference using Kaplan-Meier estimator and log-rank test [11]. The 41 DRFs are used as input to a randrom forest (RF) classifier to separate rGBM patients into groups corresponding to short-term survival (i.e., below the median survival time) and long-term survival (i.e., above or equal to the median survival time). As a validation step, the statistical significance of resulted patient groupings is measured using the log-rank test and Kaplan-Meier estimator. All image processing, array calculations, significance tests and classifications were performed in MATLAB R2018b.

2.1 Datasets

This study uses a dataset of 100 rGBM patients with post-contrast T1-weighted (T1-WI) MR images. All images are derived from a unique site, and acquired using the same scanner model, pixel spacing and slice thickness. The volume datasets are resampled with a common voxel resolution of 1 mm^3, for a total size of 256 × 256 × 256 voxels. We normalized the intensities within each volume to a range of 256 gray levels. ROIs were manually labelled by experts using the 3D Slicer software 3.6. The labelling was performed slice by slice without prior clinical information.

2.2 Proposed Deep Radiomic Features (DRFs)

To compute DRF, we used a pretrained 3D CNN architecture comprised of 4 layers, as shown in Fig. 1. The 3D CNN architecture details are as follows. *Input*: image size = 256 × 256 × 256 voxels. *Layer 1*: filter size = 2 × 2 × 2; stride = 2; filters = 10; Max pooling; Rectified Linear Unit (ReLU); dropout = 0.8; output = 10 feature maps of size ($64 \times 64 \times 64$). Layer 2: filter size = $2 \times 2 \times 2$; stride = 2; filters = 10; Max pooling; ReLU; dropout = 0.8; output = 10 feature maps of size ($32 \times 32 \times 32$). Layer 3: fully connected layer; output = vector size 128. Layer 4: softmax = vector size 2. The CNN was pretrained on multi-site 3D MRI data for the classification of Alzheimer's[1], using cross-entropy loss, stochastic gradient descent optimization with momentum of 0.9 and learning rate of 0.0005. The main goal of using pretrained 3D CNN is to generate multiscale texture descriptors (Fig. 1). These feature maps could be used directly as input to the classifier model (e.g., SVM, RF, etc.), however this leads to overfitting since the number of features largely exceeds the number of training samples. Instead, we apply conventional functions (e.g., Haralick's features [12]) to quantify the structure/texture within feature maps of deep CNN layers. Considering only the 3D feature maps in layer 1 (10 feature maps) and layer 2 (10 feature maps), we compute 41 DRF derived from the histogram, GLCM, NGTDM and GLZSM of 20 feature maps, respectively as following: (1) Histogram features (mean, variance, skewness, kurtosis, energy and entropy) encode the intensity level distribution for the image in the feature maps; (2) For GLCM, NGTDM and GLZSM features: image intensities were uniformly resampled to 32 grey-levels, averaging all the 19 GLCM features (angular second moment, contrast, correlation, sum of squares variance, homogeneity, sum average, sum variance, sum entropy, entropy, difference variance, difference entropy, information correlation$_1$, information correlation$_2$, autocorrelation, dissimilarity, cluster shape, cluster prominence, maximum probability and inverse difference [12]) across 52 GLCMs derived from 13 angles and 4 offsets, 5 NGTDM features (coarseness, contrast, busyness, complexity and texture strength [13]), and 11 GLZSM features (small zone size emphasis, large zone size emphasis, low gray-level zone emphasis, high gray-level zone emphasis, small zone/low gray emphasis, small zone/high gray emphasis, large zone/low gray emphasis, large zone/high gray emphasis, gray-level non-uniformity, zone size non-uniformity and zone size percentage [14]). Averaging the 41 features across the 20 feature maps is considered as the final descriptors, called deep radiomic features-DRF. Similarly, we computed the 41 SRF directly from original ROI images.

2.3   Classifications and Survival Analysis

To assess the proposed DRF in survival analysis, we considered days to death (i.e., censorship = 1) or days to last visit (i.e., censorship = 0) in uni- and multi-variate analyses. For the 41 DRF continuous features, the median value was used as threshold value to separate patients into two groups. For each group, the Kaplan-Meier method was considered to describe the time-to-event distributions for each feature. The log-rank significance test was then employed to assess if either group was associated with the incidence of an event. An event/censor was defined as death or the last patient visit. To account for the multiple significance tests (41×2 variables), we corrected the p values Holm-Bonferroni method [15] and considered features with corrected $p < 0.05$ as significant. A multivariate analysis based on RF model was conducted on all the patient datasets (n = 100).

---

[1] https://www.github.com/hagaygarty/mdCNN.

Uncensored patients (n = 6, alive patients) were included and were assigned the average survival time of the remaining patients with a time-to-death greater or equal to their own, as of the time at the last visit. Patients were then grouped into either short-term or long-term survivors. The threshold value dividing these two groups was their median survival of 14.88 months. We then used the combined 41 DRF as input for RF classifier model in a 5-fold cross-validation strategy. We used the RF classifier with 500 trees to predict the short- term and long-term survival outcome. The performance value is computed as the average AUC obtained across all 5 folds. To compare the DRF with the SRF we applied a similar 5-fold cross-validation strategy using the 41 radiomic features that were computed directly from the ROIs. To compare AUC value derived from DRF and SRF, we calculated significance using the chi-square test [16]. Importance values of the various features were computed within every RF tree, then averaged over the entire ensemble and normalized by dividing them by the ensemble's standard deviation. Positive importance values were considered predictive for an event.

3 Results

Figure 2a shows heatmaps of log-rank test p-values (negative $\log_{10}$ scale) for groups of patients divided by the median value of features. One feature derived from DRF (*High Gray-Level Zone Emphasis*) is associated ($p < 0.01$) with survival outcome of rGBM patients. In general, DRFs show a greater relationship to survival outcome than SRFs. As shown in Fig. 2b, longer survival was associated with a lower *High Gray-Level Zone Emphasis* value (HR = 2.28; CI = 1.46 − 3.57; 13.36 vs. 16.45 months). Notably, *High Gray-Level Zone Emphasis* describes the heterogeneity texture of GBM tumor.

Assessment of the accuracy of RF models using the DRF or SRF as input features is done in Fig. 2c. DRFs lead to a significantly higher accuracy ($p < 0.05$), with an average AUC of 89.15% compared to 78.07% using SRF, for predicting the short-term and long-term of survival outcome of rGBM patients. Comparing the AUCs of the predicted groups using the DRF and SRF, Chi-square test showed a significant p value < 0.0001. This result is consistent with the previous finding using the univariate analysis, in which DRF is more relevant than the SRF in predicting the survival. Once again, we applied the Kaplan-Meier estimator and log-rank test on the predicted groups (short-term and long-term survival) obtained by the RF classifier (Fig. 2d). We observe that the patient groups obtained by DRF or SRF have significantly different survival outcomes with $p = 1.5 \times 10^{-6}$, HR = 2.9, CI = 1.82 − 4.7 and $p = 6.8 \times 10^{-6}$, HR = 2.96, CI = 1.86 − 4.69, respectively. To assess the importance of individual features, we combined the DRF and SRF to train the RF classifier model (Fig. 2e). We find that the most predictive features (importance features > 0) are from DRF (i.e., Large Zone/Low Gray Emphasis).

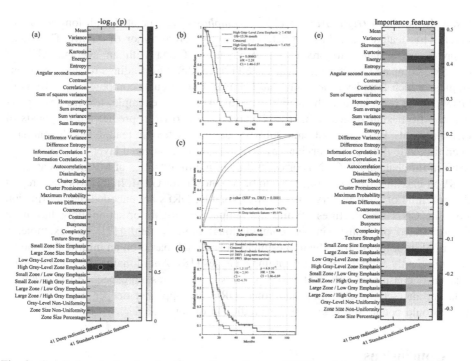

Fig. 2. Survival analysis using the deep radiomic features (DRF) and standard radiomic features (SRF). (a) Heatmap of log-rank test p values (negative $\log_{10}$ scale) of survival difference between separated by individual features. Significant features (i.e., corrected $p < 0.05$) are indicated with a black-green circle. (b) Kaplan-Meier survival curves obtained for only significant feature (High Gray-Level Zone Emphasis). (c) The area under the ROC curves (average AUC on 5 folds) obtained by the RF classifier using DRF and SRF, for predicting patients with short-term (below-median) or long-term (above-median survival) survival outcome. (d) Kaplan-Meier curves of rGBM patients that significantly predicted by RF classifier model using DRF and SRF. Solid curves correspond to the long survival group and dot curves to the short survival group. (e) Importance of individual features for predicting the survival group with the RF classifier. Positive and negative values correspond to predictive and non-predictive features, respectively.

4 Discussion

Most models based on the radiomic analysis for GBM use SRFs which include histogram, texture, and shape features derived from MR images as a non-invasive means for predicting tasks [4, 5]. SRFs have been established as a technique for quantifying the heterogeneity related to tissue abnormalities [4, 5]. Furthermore, these studies have proven the link between imaging features and clinical outcomes. Deep multi-CNN channels corresponding to MRI modalities with SVM model have been recently demonstrated to effectively predict patient survival [8].

This study proposed an objective framework using DRF based on pretrained 3D CNN with RF classifier to predict the survival outcome of rGBM. We showed that the low value of deep *High Gray-Level Zone Emphasis* is associated with long-term survival. This feature describes the heterogeneity of active tumor of rGBM. The DRFs improve performance to predict the survival of rGBM with an average AUC of 89.15%, compared to 78.07% using the STF (Fig. 2c). Our findings relate to previous studies, which found various radiomic features, in particular computed from ROIs of GBM, to be associated with overall survival [4, 5]. For example, texture descriptors derived from joint intensity matrix [4] have been shown to predict the prognosis of GBM patients. Likewise, radiomic subtypes defined by texture feature enrichment have been linked to differential survival [5]. Recently, fully-connected layer features derived from multi-channel 3D CNNs corresponding to multimodalities of MR images as input to SVM classifier were demonstrated to be prognostic for GBM. In this work, we report the potential of DRF in survival analysis. Notably, DRFs are encoded the information flow in CNN architectures. While, the impact of the information flow has been demonstrated previously in the computational biology field [17]. For example, a maximal information transduction estimation approach based on an information model was efficiently applied for transcriptome analyses [18]. Additionally, the findings from this study considered only 100 rGBM patients that requires a prospective validation on a larger dataset. Investigating additional DRFs by applying deeper architectures of CNN could potentially lead to better prediction of survival.

5   Conclusions

We proposed deep radiomic features derived from a 3D CNN to predict the survival of recurrent GBM patients. Our results show that these features lead to a higher classification performance than the standard radiomic features. Involving multi MRI modalities of rGBM tumor and more complex 3D CNN architectures to combine deep features with multi-omics (e.g., genetics + transcriptomics) could hold great promise for predicting clinical outcomes in rGBM patients.

References

1. Stupp, R., et al.: Changing paradigms—an update on the multidisciplinary management of malignant glioma. Oncologist **11**, 165–180 (2006). https://doi.org/10.1634/theoncologist.11-2-165
2. Sizoo, E.M., et al.: Measuring health-related quality of life in high-grade glioma patients at the end of life using a proxy-reported retrospective questionnaire. J. Neurooncol. **116**, 283–290 (2014). https://doi.org/10.1007/s11060-013-1289-x
3. Sharma, M., et al.: Outcomes and prognostic stratification of patients with recurrent glioblastoma treated with salvage stereotactic radiosurgery. J. Neurosurg. **1**, 1–11 (2018)
4. Chaddad, A., Daniel, P., Desrosiers, C., Toews, M., Abdulkarim, B.: Novel radiomic features based on joint intensity matrices for predicting glioblastoma patient survival time. IEEE J. Biomed. Health Inform. **23**, 795–804 (2019)
5. Rathore, S., et al.: Radiomic MRI signature reveals three distinct subtypes of glioblastoma with different clinical and molecular characteristics, offering prognostic value beyond IDH1. Sci. Rep. **8**, 5087 (2018)

6. Litjens, G., et al.: A survey on deep learning in medical image analysis. Med. Image Anal. **42**, 60–88 (2017)
7. Gu, J., et al.: Recent advances in convolutional neural networks. Pattern Recogn. **77**, 354–377 (2018)
8. Nie, D., et al.: Multi-channel 3D deep feature learning for survival time prediction of brain tumor patients using multi-modal neuroimages. Sci. Rep. **9**, 1103 (2019)
9. Chaddad, A., Desrosiers, C., Niazi, T.: Deep radiomic analysis of MRI related to Alzheimer's disease. IEEE Access **6**, 58213–58221 (2018)
10. Chaddad, A., Toews, M., Desrosiers, C., Niazi, T.: Deep radiomic analysis based on modeling information flow in convolutional neural networks. IEEE Access **7**, 97242–97252 (2019)
11. Chaddad, A., Sabri, S., Niazi, T., Abdulkarim, B.: Prediction of survival with multi-scale radiomic analysis in glioblastoma patients. Med. Biol. Eng. Comput. **56**, 2287–2300 (2018)
12. Haralick, R.M.: Statistical and structural approaches to texture. Proc. IEEE **67**, 786–804 (1979)
13. Amadasun, M., King, R.: Textural features corresponding to textural properties. IEEE Trans. Syst. Man Cybern. **19**, 1264–1274 (1989)
14. Thibault, G., et al.: Texture indexes and gray level size zone matrix application to cell nuclei classification (2009)
15. Holm, S.: A simple sequentially rejective multiple test procedure. Scand. J. Stat. **6**(2), 65–70 (1979)
16. Sumi, N.S., Islam, M.A., Hossain, M.A.: Evaluation and computation of diagnostic tests: a simple alternative. Bull. Malays. Math. Sci. Soc. **37**, 411–423 (2014)
17. Deng, Y., Bao, F., Deng, X., Wang, R., Kong, Y., Dai, Q.: Deep and structured robust information theoretic learning for image analysis. IEEE Trans. Image Process. **25**, 4209–4221 (2016)
18. Deng, Y., et al.: Information transduction capacity reduces the uncertainties in annotation-free isoform discovery and quantification. Nucleic Acids Res. **45**, e143 (2017)

cuRadiomics: A GPU-Based Radiomics Feature Extraction Toolkit

Yining Jiao[1], Oihane Mayo Ijurra[1], Lichi Zhang[1], Dinggang Shen[2], and Qian Wang[1(✉)]

[1] Institute for Medical Imaging Technology, School of Biomedical Engineering,
Shanghai Jiao Tong University, Shanghai, China
wang.qian@sjtu.edu.cn
[2] Department of Radiology and BRIC,
University of North Carolina, Chapel Hill, USA

Abstract. Radiomics is widely-used in imaging based clinical studies as a way of extracting high-throughput image descriptors. However, current tools for extracting radiomics features are generally run on CPU only, which leads to large time consumption in situations such as large datasets or complicated task/method verifications. To address this limitation, we have developed a GPU based toolkit namely cuRadiomics, where the computing time can be significantly reduced. In cuRadiomics, the CUDA-based feature extraction process for two different classes of radiomics features, including 18 first-order features based on intensity histograms and 23 texture features based on gray level cooccurrence matrix (GLCM), has been developed. We have demonstrated the advantage of the cuRadiomics toolkit over CPU-based feature extraction methods using BraTS18 and KiTS19 datasets. For example, regarding the whole image as ROI, feature extraction process using cuRadiomics is 143.13 times faster than that using PyRadiomics. Thus, the potential advantage provided by cuRadiomics enables the radiomics related statistical methods more adaptive and convenient to use than before. Our proposed cuRadiomics toolkit is now publicly available at https://github.com/jiaoyining/cuRadiomics.

Keywords: Radiomics · CUDA · Feature extraction · GPU

1 Introduction

Radiomics is an advanced way of extracting high-throughput features. It can transform images into mineable data, which facilitates the understanding of different data patterns that could not be appreciated by human eye alone [1]. Recently, medical images such as computed tomography (CT), magnetic resonance (MR), or positron emission tomography (PET) images have gained crucial importance of mirroring pathology of patients. Therefore, radiomics features are increasingly used in clinical decision-

This research was supported by the grants from the National Key Research and Development Program of China (No. 2017YFC0107602 and No. 2018YFC0116400), Medical Engineering Cross Research Foundation of Shanghai Jiao Tong University (ZH2018QNA67).

H. Mohy-ud-Din and S. Rathore (Eds.): RNO-AI 2019, LNCS 11991, pp. 44–52, 2020.
https://doi.org/10.1007/978-3-030-40124-5_5

making tasks for their strong predictive power and promising interpretability, such as the diagnosis/prognosis of different diseases and the design of personalized treatments. For example, the effectiveness of radiomics features is substantiated in many oncology researches including lung cancer [2], colorectal cancer [3], breast cancer [4] and glioblastoma [5].

Generally, the pipeline of medical image analysis involves the following major steps: (1) collect medical images satisfying certain recruiting criterion; (2) outline regions of interest (ROIs) manually or automatically; (3) extract quantitative radiomics features within the ROIs; (4) apply data mining strategies such as statistical analysis and/or machine learning models to explore data patterns; (5) validate the performance of the proposed model. For feature extraction process, various hand-crafted features are designed with the intent of serving as image descriptors of medical images. For example, first-order features based on histograms and high-order texture features, such as Haralick features [6] based on gray level cooccurrence matrix (GLCM) are the features that could be extracted for medical purposes. Furthermore, many toolkits are developed to extract these radiomics features. Among them, PyRadiomics toolkit [7], which is developed in C and Python languages and can only be run with CPU support, is the most widely used one.

However, there are some limitations in current radiomics feature extraction toolkits which hamper their clinical applications. Firstly, these algorithms are usually designed in a serialized manner, in which the images are read and analyzed one by one. Secondly, the calculation of the intermediate matrices such as GLCM, consisting in reading and processing voxels in images subsequently, is largely run on CPU. Thus, current feature extraction methods are very time consuming and inconvenient, exacerbating the complexity of many clinical problems especially when involving large datasets.

Efforts are made to resolve the problem of extracting GLCM on large, sparse microscopy images using CUDA, by iterating over multi-cell images with a step width to compute GLCM and Haralick features, neglecting zero values of GLCM and rearranging the computing order of features [8]. In this way, one multi-cell image is split into several parts to make the in-part computation parallelized and the out-part computation iterative. However, the iterative operation is still time consuming, especially for images that are not sparse like MR and CT images. Besides, the benefit of this method is limited since the extracted features' number is small and the images are still processed one by one, instead of a batch style.

To address these issues in current feature extraction methods, in this paper we develop a GPU-based radiomics feature extraction toolkit, called cuRadiomics using the CUDA computation platform. Specifically, the cuRadiomics intends to extract 41 radiomics features which can be divided in two groups: (1) 18 first-order features based on the intensity histogram and (2) 23 texture features based on GLCM. Furthermore, the cuRadiomics provides several potential advantages when compared with conventional CPU-based implementation. First, radiomics features are batch-processed; Second, each value of matrices like GLCM and the radiomics features except features involving maximum/minimum operation are able to be calculated simultaneously. Therefore, the cuRadiomics contribute to large-fold acceleration of feature extraction. The proposed toolkit, not only enables the acceleration of the current radiomics analysis pipelines, can also be seamlessly merged into other deep learning networks such as deep reinforcement learning, to make some heuristic searches for lesions or tumors.

2 Methods

As previously mentioned, the developed cuRadiomics toolkit involves two classes of radiomics features, i.e., (1) 18 first-order features based on intensity histogram and (2) 23 texture features based on GLCM. The feature extraction pipeline is basically divided into three major steps: (1) calculation of the histograms or feature matrices; (2) calculation of vectors representing the properties of the obtained matrices; (3) calculation of the radiomics features using matrices and properties obtained from the above two steps. The workflow of cuRadiomics is presented in Fig. 1. Note that we set the background voxels in the label map as −1, so that the cuRadiomics can focus only on the annotated ROIs.

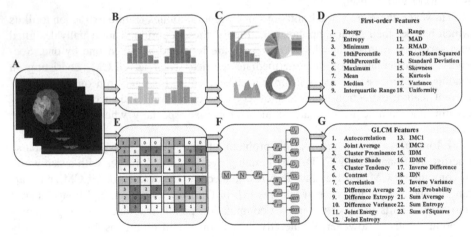

Fig. 1. The workflow of cuRadiomics. A: MR slices as the input of cuRadiomics. B: Histograms of the input MR slices. C: Properties calculated from the histograms. D: 18 first-order features of input images. E: GLCMs of the MR slices. F: Properties calculated from the GLCMs, from left to right showing the time order of the calculation for GLCM properties. G: 23 texture features based on GLCM. **MAD:** Mean Absolute Deviation. **RMAD:** Robust Mean Absolute Deviation. **IMC1:** Informational Measure of Correlation 1. **IMC2:** Informational Measure of Correlation 2. **IDM:** Inverse Difference Moment. **IDMN:** Inverse Difference Moment Normalized. **IDN:** Inverse Difference Normalized.

2.1 Computation of First-Order Features

The pipeline for the extraction of first-order features is shown in Fig. 1(B, C, D). Compared to CPU-based implementation, images are batch processed. The calculation of the histogram is optimized by counting voxels of each grayscale simultaneously, whereas, in PyRadiomics, the voxels are accessed one by one and counted according to their grayscales. Then, the properties of histograms, such as cumulative distribution functions and the numbers of voxels within ROIs, are calculated to extract the 18 first-order features.

2.2   Computation of Texture Features

The pipeline of texture feature extraction based on GLCM is shown in Fig. 1(E, F, G). Different with PyRadiomics, here the images are batched as input. For GLCM, the number of a certain kind of gray-level pairs is the value of the corresponding element in GLCM. In cuRadiomics, each element of the GLCMs is calculated in an individual thread independently, and the voxels within ROIs can be accessed simultaneously to count the number of all kinds of gray-level pairs within the ROIs. In this way, each element of GLCM can be calculated simultaneously. However, in PyRadiomics, the voxels need to be read one by one to count the number of each kind of gray-level pairs, resulting in great time consumption, especially for large ROIs. The properties of GLCM are more complicated than those of histogram, therefore we classify them according to their dependence on each other, and the calculation is made in four steps as shown in Fig. 1(F). Then, the GLCM texture features can be derived simultaneously from the above calculated properties within one CUDA kernel function.

3   Quantitative Evaluation

In this section, we evaluate the performance of the proposed cuRadiomics toolkit and the widely-used PyRadiomics (2.1.2), by applying them to extract radiomics features on two different medical datasets.

3.1   Environment

We perform experiments of our proposed cuRadiomics and the PyRadiomics toolkits using a system with an Intel i7-7700K Processor, containing 4 cores (8 with hyper threading) and 64 GB of DDR4 RAM. The GPU in use on this system is the NVIDIA GeForce GTX 1080Ti edition GPU, with 3584 CUDA cores clocked at 1.6 GHz and 11 GB of GDDR5 memory. It supports a maximum memory speed of 11 Gbps, with a maximum memory bandwidth of 484 Gbps. The GP102chip (1080Ti) has 2816 Kb of L2 cache, and 28 Streaming Multiprocessors, with a theoretical peak performance of 11.5 TFLOPs for single precision floating point operations. The CPU platform has a theoretical peak performance of 268.8 GFLOPs.

3.2   Dataset

Two different medical datasets are used to compare the performances of our proposed GPU-based cuRadiomics toolkit and the PyRadiomics, a CPU-based radiomics feature extraction toolkit:

- MICCAI 2018 BraTS training dataset [9, 10]. Brain MR images of 285 subjects are provided with Flair, T1, T1CE and T2 modalities, as well as their respective segmented enhancing tumor (ET), the peritumoral edema (ED), and the necrotic and non-enhancing tumor core (NCR/NET). The size of each slice is 240 × 240.
- KiTS19 dataset [11]. Totally 210 CT images are provided with the segmented kidneys and tumors. For each slice of the CT images, its size is 512 × 512.

3.3 Experimental Results

Note that to make fair comparison of the time consumption of cuRadiomics and PyRadiomics, the cuRadiomics toolkit is packaged into a dynamic link library, therefore it can be called directly in python platform where the images are read and preprocessed. For the radiomics feature extraction process, the medical images and masks are initially read in Python platform, then we use cuRadiomics and PyRadiomics toolkits to extract the radiomics features.

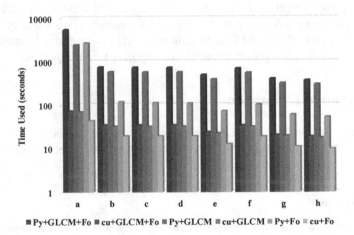

Fig. 2. Performance of CuRadiomics and PyRadiomics on BraTS18 dataset. Py+CLCM+Fo: Time used by PyRadiomics for extracting GLCM features and first-order features. cu+GLCM +Fo: Time used by cuRadiomics for extracting GLCM features and first-order features. Py +GLCM: Time used by PyRadiomics for extracting GLCM features. cu+GLCM: Time used by cuRadiomics for extracting GLCM features. Py+Fo: Time used by PyRadiomics for extracting first-order features. cu+Fo: Time used by cuRadiomics for extracting first-order features. Each group of bars shows the comparison of time used for calculating features in different ROIs. a: the whole image; b: ED combined with ET and NCR/NET; c: ED combined with NCR/NET; d: ED combined with ET; e: ET combined with NCR/NET; f: ED; g: NCR/NET; h: ET.

Table 1. Ratio of time used by PyRadiomics and cuRadiomics on BraTS18 dataset. (1) Extracting GLCM features and first-order features; (2) Extracting GLCM features; (3) Extracting first-order features.

Time	a	b	c	d	e	f	g	h
(1)	74.62	21.53	21.15	21.13	20.78	20.50	19.83	19.93
(2)	35.06	18.15	17.99	17.87	17.59	17.51	16.31	16.93
(3)	61.80	6.12	5.82	5.68	5.77	5.49	5.44	5.41

For BraTS2018 dataset, we get eight types of ROIs from the given segmentation, i.e., a: the whole image; b: ED combined with ET and NCR/NET; c: ED combined with NCR/NET; d: ED combined with ET; e: ET combined with NCR/NET; f: ED; g: NCR/NET; h: ET.

For KiTS2019 dataset, we get four kinds of ROIs in a similar way as BraTS2018, i.e., (a) the whole image; (b) kidney combined with tumor; (c) kidney; (d) tumor.

For each kind of ROIs in medical images, radiomics features are extracted and the total time used in each kind of ROI is recorded as shown in Figs. 2 and 3, respectively. The ratio of time used by PyRadiomics and cuRadiomics is reported in Tables 1 and 2 respectively.

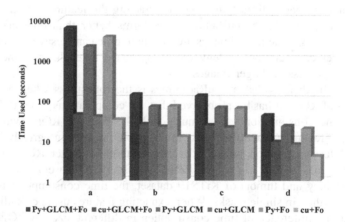

Fig. 3. Performance of cuRadiomics and PyRadiomics on KiTS19 dataset. Py+CLCM+Fo: Time used by PyRadiomics for extracting GLCM features and first-order features. cu+GLCM +Fo: Time used by cuRadiomics for extracting GLCM features and first-order features. Py +GLCM: Time used by PyRadiomics for extracting GLCM features. cu+GLCM: Time used by cuRadiomics for extracting GLCM features. Py+Fo: Time used by PyRadiomics for extracting first-order features. cu+Fo: Time used by cuRadiomics for extracting first-order features. Each pair of bars shows the comparison of time used for calculating features in different ROIs. a: the whole image; b: kidney combined with tumor; c: kidney; d: tumor.

Table 2. Ratio of time used by PyRadiomics and cuRadiomics on KiTS19 dataset. (1) Extracting GLCM features and first-order features; (2) Extracting GLCM features; (3) Extracting first-order features.

Time	a	b	c	d
(1)	143.13	5.45	5.42	4.52
(2)	55.02	3.16	3.05	2.84
(3)	114.03	5.64	5.10	4.79

4   Discussion

As presented in Figs. 2 and 3, it can be observed that the time consumed using cuRadiomics for extracting the radiomics features on two medical datasets in all types of ROIs is dramatically decreased, compared with PyRadiomics. Furthermore, when extracting both first-order and GLCM features, the cuRadiomics is 74.62 times faster than PyRadiomics on BraTS18 dataset while 143.13 times faster than PyRadiomics on KiTS19 dataset. Thus, the effectiveness of our proposed cuRadiomics toolkit is demonstrated, making it desirable for fast radiomics feature extraction.

The most obvious improvement appears when the radiomics features are extracted on the whole image, because, compared with other kinds of ROIs (that do not necessarily exist on every slice), all slices are imported to the cuRadiomics toolkit to make more computations parallelized. Also, when extracting the features on KiTS19 using "the whole image" mask, the cuRadiomics performs better than on BraTS18. This suggests that the larger the input images are, the more time can be saved, for the reason that the parallel computation strategy of "using space to exchange time" is more effective when processing larger images.

Another strength provided by cuRadiomics is the robustness when dealing with different sizes of ROIs. It has been observed that, for computations using PyRadiomics toolkit, the time consumption varies much with the sizes of ROIs, with the time consumption of feature extraction process in the whole image being greatly larger than in smaller ROIs. In addition, when extracting features in larger ROIs such as the combined three masks (ED, ET, and NCR/NET) of BraTS18 dataset and the combined two masks (kidney and tumor) of KiTS19 dataset, the time consumption is also significantly larger than in single masks. When extracting features using cuRadiomics, we observed that the difference of time consumption in different sizes of ROIs is not so large as in Pyradiomics. The slight variation of time consumption in cuRadiomics is due to the time used in operations such as coping larger arrays from CPU to GPU. However, this additional time cost of copying array is limited because images are batch processed and imported into cuRadiomics, making the copying operation less frequent. Therefore, the sizes of the masks do not affect much the performance of our developed cuRadiomics compared to PyRadiomics. This strength enables cuRadiomics suitable for high-resolution images and to accomplish some heuristic search and detection tasks.

It is also noted that in the case of feature extraction process on relatively small masks, the strength is relatively not so outstanding compared with bigger masks. When extracting features of ET on BraTS18 dataset, the cuRadiomics is 19.93 times faster than PyRadiomics. The results also show that the extraction of features from tumors on KiTS19 dataset is only 4.52 times faster than using cuRadiomics. This is quite understandable because, in hundreds of slices of a CT image in KiTS19 dataset, only a few slices have tumors, leading to less computations being parallelized.

In summary, our proposed method enables faster extraction of high-throughput and quantitative radiomics features, which makes the traditional radiomics research more efficient, especially in the large dataset. Furthermore, due to its advantages in optimized algorithm and parallelized processing, our proposed cuRadiomics can be embedded into deep learning platform such as TensorFlow and PyTorch to bring the inter-pretability of radiomics features, for which deep learning networks cannot provide. This property is very beneficial for many clinical applications such as lesion detection and tumor segmentation.

5 Conclusion

In this paper, a GPU-based toolkit, namely cuRadiomics, has been developed to enable parallelized radiomics feature extraction. In this toolkit, we developed parallelized extraction of both 18 first-order features and 23 texture features based on GLCM. Compared with CPU-based feature extraction toolkits, such as PyRadiomics, our proposed cuRadioimics is dramatically faster, allowing for more efficient research based on radiomics. Moreover, cuRadiomics can be embedded in to more complicated methods or tasks, such as deep learning methods, to help radiologists in different clinical applications. Our proposed cuRadiomics has been tested with 2D slices, to extract a total of 41 radiomics features. In our future works, we will continue devel-oping cuRadiomics to expand its functions in terms of more additional features, as well as to enable faster feature extraction in the 3D volumes.

References

1. Lambin, P., et al.: Radiomics: the bridge between medical imaging and personalized medicine. Nat. Rev. Clin. Oncol. **14**, 749 (2017)
2. Aerts, H.J., et al.: Decoding tumour phenotype by noninvasive imaging using a quantitative radiomics approach. Nat. Commun. **5**, 4006 (2014)
3. Huang, Y.Q., et al.: Development and validation of a radiomics nomogram for preoperative prediction of lymph node metastasis in colorectal cancer. J. Clin. Oncol. **34**, 2157–2164 (2016)
4. Li, H., et al.: Quantitative MRI radiomics in the prediction of molecular classifications of breast cancer subtypes in the TCGA/TCIA data set. NPJ Breast Cancer **2**, 1–10 (2016)
5. Gevaert, O., et al.: Glioblastoma multiforme: exploratory radiogenomic analysis by using quantitative image features. Radiology **273**, 168–174 (2014)
6. Haralick, R.M., Shanmugam, K., Dinstein, I.: Textural features for image classification. IEEE Trans. Syst. Man Cybern. **3**, 610–621 (1973)
7. van Griethuysen, J.J., et al.: Computational radiomics system to decode the radiographic phenotype. Cancer Res **77**, e104–e107 (2017)

8. Gipp, M., et al.: Haralick's texture features computation accelerated by GPUs for biological applications. In: Bock, H., Hoang, X., Rannacher, R., Schlöder, J. (eds.) Modeling, Simulation and Optimization of Complex Processes. Springer, Heidelberg (2012). https://doi.org/10.1007/978-3-642-25707-0_11

9. Bakas, S., et al.: Advancing the cancer genome atlas glioma MRI collections with expert segmentation labels and radiomic features. Sci. Data **4**, 170117 (2017)

10. Menze, B.H., et al.: The multimodal brain tumor image segmentation benchmark (BRATS). IEEE Trans. Med. Imaging **34**, 1993–2024 (2015)

11. Heller, N., Sathianathen, N., Kalapara, A., et al.: The KiTS19 challenge data: 300 kidney tumor cases with clinical context, CT semantic segmentations, and surgical outcomes (2019). arXiv:1904.00445

On Validating Multimodal MRI Based Stratification of IDH Genotype in High Grade Gliomas Using CNNs and Its Comparison to Radiomics

Tanay Chougule[1], Sumeet Shinde[1], Vani Santosh[2], Jitender Saini[2], and Madhura Ingalhalikar[1(✉)]

[1] Symbiosis Center for Medical Image Analysis,
Symbiosis International University, Pune 412115, India
head@scmia.edu.in
[2] National Institute of Mental Health and Neurosciences,
Bengaluru 560029, India

Abstract. Radiomics based multi-variate models and state-of-art convolutional neural networks (CNNs) have demonstrated their usefulness for predicting IDH genotype in gliomas from multi-modal brain MRI images. However, it is not yet clear on how well these models can adapt to unseen datasets scanned on various MRI scanners with diverse scanning protocols. Further, gaining insight into the imaging features and regions that are responsible for the delineation of the genotype is crucial for clinical explainability. Existing multi-variate models on radiomics can provide the underlying signatures while the CNNs, despite better accuracies, more-or-less act as a black-box model. This work addresses these concerns by training radiomics based classifier as well as CNN classifier with class activation mapping (CAMs) on 147 subjects from TCIA and tests these classifiers directly and through transfer learning on locally acquired datasets. Results demonstrate higher adaptability of Radiomics with average accuracy of 75.4% than CNNs (68.8%), however CNNs with transfer learning demonstrate superior predictability with an average accuracy of 81%. Moreover, our CAMs display precise discriminative location on various modalities that is particularly important for clinical interpretability and can be used in targeted therapy.

Keywords: Class activation map (CAM) · Convolutional neural nets (CNN) · Radiomics · IDH · Gliomas

1 Introduction

Gliomas are the most common type of neoplasms. The 2016 World Health Organization (WHO) classification of brain tumors recognized several new entities based on genotypes in addition to histological phenotypes [1]. Amongst these, mutations in isocitrate dehydrogenase 1(IDH1) were considered crucial as these are associated with longer overall survival [2]. Currently, the IDH genotype is identified via immuno-histochemical analysis following biopsy or surgical resection. Therefore, developing

H. Mohy-ud-Din and S. Rathore (Eds.): RNO-AI 2019, LNCS 11991, pp. 53–60, 2020.
https://doi.org/10.1007/978-3-030-40124-5_6

non-invasive pre-operative markers for IDH genotype is clinically important as it can not only aid prognosis but also support treatment planning and therapeutic intervention.

Growing evidence has revealed the feasibility of multi-modal MRI images to probe the fundamental tumor type by quantifying the underlying imaging traits. Radiomics, a recently developed high throughput approach has demonstrated potential in characterizing IDH genotype [3]. It quantifies various patterns, textures and shapes of the tumoral region on the MRI images and when used in a multi-variate classification framework can predict the outcomes- in this case the IDH mutation status of the glioma. Multiple studies have demonstrated the clinical implication and utility of radiomics based genotype prediction especially on multimodal MRI where the combination of features extracted from multiple modalities that include gadolinium enhanced T1-weighted (T1ce), T2-weighted, and fluid attenuation inversion recovery (FLAIR) boost the predictive outcomes [4, 5]. Moreover, the most discriminative textural and shape features can also be extracted that facilitate the underlying signatures for the genotype. However, the procedures involving radiomics can be laborious at it involves multiple processing steps such as intensity normalization, tumor segmentation, feature extraction and selection and multi-variate classification using models such as support vector machines [6], random forests [7] etc. To reduce the complications involved, recent studies have employed deep neural networks such as convolutional neural nets (CNNs) [8] that operate directly on the images and eliminate manual feature extraction and selection steps. Studies using CNNs have demonstrated higher accuracies in delineating gliomas with IDH mutation from IDH wildtype [9, 10] on large datasets. Despite the promising results CNNs do not provide insights into the regions or features that discriminate one class from another which is crucial for clinical explainability. Moreover, all the studies to this date, although use data from multiple sites, do not perform leave-one-site-out type of analysis creating uncertainties about the adaptability to unseen datasets acquired from different scanners with diverse scanning protocols. To mitigate the aforementioned issues, this work trains radiomics based and CNN classifier on a large open source dataset and tests it on locally acquired datasets to assess the applicability of these models. For clinical interpretability, we extract the underlying discriminative radiomics features and for the CNN model we employ a high-resolution class activation map (HR-CAM) technique to demonstrate the regions of discrimination on multiple modalities.

2   Method

2.1   Study Cohort and Imaging

Our datasets consisted of a clinical cohort of subjects that had undergone surgical resection and standard post-surgical care and were identified retrospectively after reviewing the medical records. IDH mutation status was determined after resecting the tumor, via immuno-histochemistry or next generation sequencing. Our first cohort included 23 patients (age 38.39 ± 13.16 yrs, M:F 17:6) with grade III and IV glioblastomas with IDH mutation and 14 patients (age 43.78 ± 16.76, M:F 5:9) with grade III and IV glioblastoma without the mutation while the second cohort included

14 patients (age 43.64 ± 11.25yrs, M:F 7:7) with grade III and IV glioblastomas with IDH mutation and 10 patients (age 53.3 ± 12.16, M:F 5:5) with grade III and IV glioblastoma without the mutation in IDH. Cohort 1 was scanned on a Philips Achieva 3T MRI scanner where the T1 weighted (T1ce) was acquired using TR/TE of 8.7/3.1 misusing a TFE sequence, fluid attenuation inversion recovery (FLAIR) were acquired using TR/TE/T1 of 11000/125/2800 with in plane resolution of 0.5 × 0.5 mm and T2 weighted imaging was performed using TR/TE = 3600/80 ms and 0.5*0.5 mm resolution in the axial plane. For cohort 2, T1ce scans were obtained using TR/TE = 2200/2.3 using T1 MPRAGE sequence with 1 * 1 * 1 mm isotropic resolution. T2 protocol consisted of TR/TE ranging from 5500/90 ms and 0.5 * 0.5 mm resolution in the axial plane. FLAIR images were acquired using similar parameters as cohort 1. The dataset used for training the classifiers was taken from TCIA/TCGA with 90 subjects with IDH mutation and 57 wildtype and were scanned on multiple scanners. The details of the dataset can be found in Bakas et al. [11].

2.2   Image Processing

The TCIA training set was processed using multiple steps. All MRI volumes were co-registered to a T1 template using affine registration through the Linear Image Registration Tool (FLIRT) of FMRIB Software Library (FSL) [12], and resampled to 1 mm^3 voxel resolution. The volumes of all the modalities for each patient were then skull-stripped using the Brain Extraction Tool (BET) from the FSL. Details of the image preprocessing can be found in Bakas et al. [11].

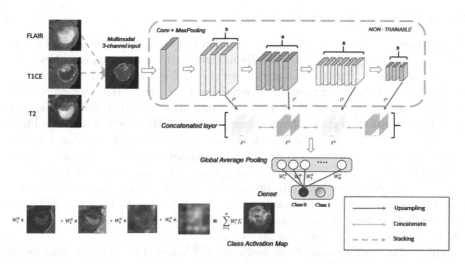

Fig. 1. CNN architecture employed for predicting the IDH genotype as well as generating HR-CAMs.

A similar process was used for our local datasets, where T2-FLAIR and T2 weighted images were first registered to the subject's T1ce image using affine registration with mutual information similarity metric using ANTs [13]. Then, brain extraction was performed using BET, followed by non-parametric, non-uniform intensity normalization to eliminate low frequency noise. Intensity normalization was performed to scale the imaging intensities into standardized ranges for each imaging modality among all subjects.

2.3   Radiomics

For the local test sets, the region of interest (ROI) covering the total tumor volumes (including the contrast enhancing, edema, and necrotic regions) was identified through a semiautomatic segmentation process. A convolutional auto-encoder with 14 layers was trained to identify edema regions from T2 and FLAIR images, and another model with same architecture was trained to identify Enhancing and Necrotic regions from T1ce images. The images and segmentation labels from BRATS-2018 (https://www.med.upenn.edu/sbia/brats2018/data.html) were employed for training and validation (training n = 206, validation n = 52) of these auto-encoders. The ROI masks predicted by the model were then corrected manually for each subject and cross-checked by an experienced neuro-radiologist. Radiomics feature extraction was performed using PyRadiomics 2.2.0 library and included 3D shape-based features, statistical features, Gray-Level Co-occurrence Matrix (GLCM), Gray-Level dependant matrix (GLDM), Gray Level Run Length Matrix (GLRLM), Gray Level Size Zone Matrix (GLSZM) and Neighboring Gray Tone Difference Matrix (NGTDM) [14]. A total number of 107 features were extracted from each modality and 321 features overall (for 3 modalities) were computed for each subject.

2.4   CNNs with Class Activation Maps

Class activation maps provide the most discriminating regions in an image, for a given task. A good class activation maps usually indicates better generalization and less over-fitting. For this task, it means that the regions provided by these CAMs are likely to be the most discriminatory regions for determining IDH status. Thus, we employed a novel 2D-CNN architecture which is derived from ResNet50 [15], and modified it to provide high resolution class activation maps, called as HR-CAMs as shown in Fig. 1.

Along with the traditional convolutional layers, ResNets employ residual learning framework which allows training deeper convolutional networks without degradation. The model consisted of 16 residual blocks, each with 3 convolutional layers and a residual connection. The last convolutional layer acquired the most abstract markers that was given as input to global average pooling (GAP) layer which averaged across the output of the convolutional layer for both the classes resulting in 2 representations, one for each class using a softmax activation.

HR-CAMs were obtained from ResNet50 in 2 stages (Fig. 1) - The input images were first trained using ResNet50 convolutional layers with a classifier at the end. The feature maps of each layer before the pooling layer were up-sampled to input image size using linear interpolation. (2) - The resulting N feature maps were then concatenated and were given as input to a global average pooling layer that was trained to optimize the weights $(W_1, W_2..., W_N)$ while keeping the previous layers frozen. High resolution class activation maps were finally obtained from these weights using the technique given by Zhou et al. [16].

Input images for CNN models were obtained by stacking 2D axial slices from 3 modalities (FLAIR, T1ce, T2), as channels. For each of these images, a boxed region around the tumor ROI, resized to shape 128 * 128 was used as input to the CNN. Data augmentation was used as another way to avoid over-fitting and to increase the training data by many folds. This was achieved using random flips, horizontal and vertical shifts and random image rotations on input training images. Along with augmentation, the input images were normalized for each modality using mean and standard deviation from training images. Finally, to avoid over-fitting, L2 regression and dropout was used in the final dense layers.

Table 1. Table showing all the results

Dataset	Radiomics			CNN		
	Sensitivity	Specificity	Accuracy	Sensitivity	Specificity	Accuracy
TCGA-CV	0.878	0.855	86.9	0.960	0.941	**95.3**
Test (1)	0.66	0.68	**67.5**	0.6	0.72	**67.5**
Test (2)	1	0.714	**83.3**	0.7	0.714	70.1
All test (transfer learning)	–	–	–	0.83	0.8	**81.25**

2.5 Training and Testing

The TCIA/TCGA data was divided into training cohort (74 Mutant, 41 WT) and validation cohort (16 Mutant, 16 WT) and the other two datasets were used for testing. The radiomic features were min-max normalized and feature selection of 75 features was performed using a variance threshold and ANOVA f-values on the training cohort.

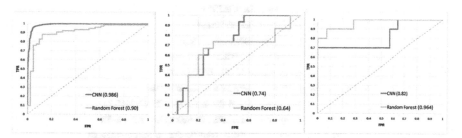

Fig. 2. ROC curves for (a) CV- on the TCGA/TCIA data (b) Local test dataset 1 and (c) Local test dataset 2.

A random forest classifier (RF Classifier) was used for classification of IDH status based on radiomics features. Hyper-parameters for the classifier were obtained using 1000 iterations of randomized search cross-validation on the training cohort. To compare the performance of the model on TCIA dataset, 5-fold cross validation was used. Finally, the classifier was trained based on the obtained hyper-parameters and tested on two local datasets to compare the adaptability of radiomics features on datasets obtained from different scanners and site.

A 2D-CNN model was trained to identify the IDH status based on multi-modal MRI images. Keras package in python was used to build and train the model. To compare the performance of the model, 5-fold cross-validation using the same dataset as RF classifier was performed. In each fold, the model was trained for 100 epochs with Adam optimizer with an initial learning rate of 0.0001 which was reduced every time the validation loss did not improve for more than 10 epochs. Finally, the model was tested on test datasets whose images normalized along modalities based on mean and standard deviation from the TCIA data.

To perform transfer learning, weights learned from the TCGA dataset were used as initial weights for the CNN and it was then trained on the combined test datasets for 100 epochs. Here we combined test cohort 1 and 2 for training and testing where 48 subjects were used for learning and 16 were used for testing.

3   Results

The cross-validation accuracy for radiomics based classifier was 86.9% and with the CNN HR-CAM it was 95.3% (Table 1). While testing the unseen data, we observed that radiomics performed with a higher accuracy (67.5% and 83.3%) while the CNN model demonstrated lower adaptability with test accuracy of 67.5% and 70.1% on dataset 1 and 2 respectively (Table 1). However, with transfer learning we could improve the performance of CNNs to 81%. Figure 2 displays the ROCs for CV and testing.

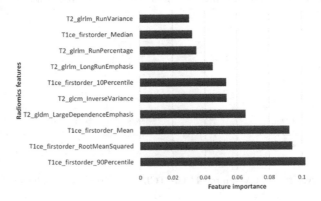

Fig. 3. Top ranked radiomics features plotted against the RF score (feature importance). It can be observed that the first order features from the enhancing area on T1ce are important in classifying IDH followed by texture features extracted on T2 images.

To understand which areas and features were significant in classifying the two groups, we computed the top 10 radiomics features that are plotted in Fig. 3. These were computed from the importance scores of the random forest classifier. It can be clearly seen that first order features from T1ce are crucial in identifying IDH genotype followed by GLCM and run length features extracted from T2-weighted images. The T2-FLAIR did not demonstrate high predictive potential. On the CNNs, we computed HR-CAMs that illustrated the most discriminative region for each subject under consideration. Figure 4 provides an example of the HR-CAMs computed on mutant case (row 1) and on wildtype (row 2). The red area on the HR-CAMs is highly weighted by the CNNs. Thus, it is significant for the delineation of the patient to that group.

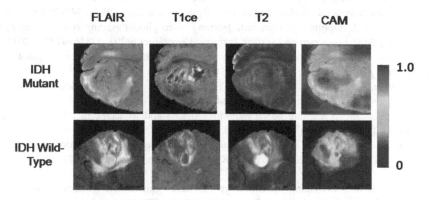

Fig. 4. Showing the HR-CAMs for (row1) –mutant case and (row2) wildtype case. The first column is the T2-FLAIR, second is the T1ce and third column in the T2 weighted image. The final column shows the HR-CAM that illustrates the most discriminative area. (Color figure online)

4   Conclusion

This work aimed at understanding the adaptability of predictive models to unseen data to classify gliomas with IDH genotype at a pre-operative radiological level. To this end, we employed Radiomics and deep nets and found that although CNNs were better in training and cross-validation, radiomics based classification was more robust on unseen data. However, CNNs with the option of using transfer learning demonstrated a boost in accuracy. Furthermore, we also demonstrated that T1ce and T2 based Radiomics features were significant in delineating IDH genotype. With the CNNs we illustrated that patient specific HR-CAMs can be employed to gain insights into the most discriminative regions that might hold implications in targeted therapy. The findings of this study are crucial as imaging prediction of IDH mutation is important and as and when IDH mutant inhibitors become clinically available, these might be used as neoadjuvant therapy.

References

1. Louis, D.N., et al.: The 2016 World Health Organization classification of tumors of the central nervous system: a summary. Acta Neuropathol. **131**(6), 803–820 (2016)
2. Houillier, C., et al.: IDH1 or IDH2 mutations predict longer survival and response to temozolomide in low-grade gliomas. Neurology **75**(17), 1560–1566 (2010)
3. Kinoshita, M., et al.: Introduction of high throughput magnetic resonance T2-weighted image texture analysis for WHO grade 2 and 3 gliomas. PLoS ONE **11**(10), e0164268 (2016)
4. Lu, C.F., et al.: Machine learning-based radiomics for molecular subtyping of gliomas. Clin. Cancer Res. **24**(18), 4429–4436 (2018)
5. Suh, C.H., et al.: Imaging prediction of isocitrate dehydrogenase (IDH) mutation in patients with glioma: a systemic review and meta-analysis. Eur. Radiol. **29**(2), 745–758 (2019)
6. Zhang, X., et al.: Optimizing a machine learning based glioma grading system using multi-parametric MRI histogram and texture features. Oncotarget **8**(29), 47816–47830 (2017)
7. Zhang, B., et al.: Multimodal MRI features predict isocitrate dehydrogenase genotype in high-grade gliomas. Neuro Oncol. **19**(1), 109–117 (2017)
8. Krizhevsky, A., Sutskever, I., Hinton, G.E.: ImageNet classification with deep convolutional neural networks. Commun. ACM **60**(6), 84–90 (2017)
9. Li, Z., et al.: Deep learning based radiomics (DLR) and its usage in noninvasive IDH1 prediction for low grade glioma. Sci. Rep. **7**(1), 5467 (2017)
10. Chang, K., et al.: Residual convolutional neural network for the determination of IDH status in low-and high-grade gliomas from MR imaging. Clin. Cancer Res. **24**(5), 1073–1081 (2018)
11. Bakas, S., et al.: Advancing the cancer genome atlas glioma MRI collections with expert segmentation labels and radiomic features. Sci. Data **4** (2017). Article number: 170117
12. Jenkinson, M., et al.: Fsl. Neuroimage **62**(2), 782–790 (2012)
13. Avants, B.B., et al.: A reproducible evaluation of ANTs similarity metric performance in brain image registration. Neuroimage **54**(3), 2033–2044 (2011)
14. Herz, C., et al.: DCMQI: an open source library for standardized communication of quantitative image analysis results using DICOM. Cancer Res. **77**(21), E87–E90 (2017)
15. He, K., Zhang, X., Ren, S., Sun, J.: Identity mappings in deep residual networks. In: Leibe, B., Matas, J., Sebe, N., Welling, M. (eds.) ECCV 2016. LNCS, vol. 9908, pp. 630–645. Springer, Cham (2016). https://doi.org/10.1007/978-3-319-46493-0_38
16. Zhou, B., et al.: Learning deep features for discriminative localization. In: CVPR (2016)

Imaging Signature of 1p/19q Co-deletion Status Derived via Machine Learning in Lower Grade Glioma

Saima Rathore[1]([⊠]), Ahmad Chaddad[2], Nadeem Haider Bukhari[3], and Tamim Niazi[2]

[1] Center for Biomedical Image Computing and Analytics,
Perelamn School of Medicine, University of Pennsylvania,
Philadelphia, PA, USA
saima.rathore@pennmedicine.upenn.edu
[2] Lady Davis Institute for Medical Research, McGill University,
Montreal, QC, Canada
[3] University of Azad Jammu and Kashmir,
Muzaffarabad, Azad Kashmir, Pakistan

Abstract. We present a new approach to quantify the co-deletion of chromosomal arms 1p/19q status in lower grade glioma (LGG). Though the surgical biopsy followed by fluorescence in-situ hybridization test is the gold standard currently to identify mutational status for diagnosis and treatment planning, there are several imaging studies to predict the same. Our study aims to determine the 1p/19q mutational status of LGG non-invasively by advanced pattern analysis using multi-parametric MRI. The publicly available dataset at TCIA was used. T1-W and T2-W MRIs of a total 159 patients with grade-II and grade-III glioma, who had biopsy proven 1p/19q status consisting either no deletion (n = 57) or co-deletion (n = 102), were used in our study. We quantified the imaging profile of these tumors by extracting diverse imaging features, including the tumor's spatial distribution pattern, volumetric, texture, and intensity distribution measures. We integrated these diverse features via support vector machines, to construct an imaging signature of 1p/19q, which was evaluated in independent discovery ($n = 85$) and validation ($n = 74$) cohorts, and compared with the 1p/19q status obtained through fluorescence in-situ hybridization test. The classification accuracy on complete, discovery and replication cohorts was 86.16%, 88.24%, and 85.14%, respectively. The classification accuracy when the model developed on training cohort was applied on unseen replication set was 82.43%. Non-invasive prediction of 1p/19q status from MRIs would allow improved treatment planning for LGG patients without the need of surgical biopsies and would also help in potentially monitoring the dynamic mutation changes during the course of the treatment.

Keywords: Low-grade gliomas · Pattern analysis · Glioblastoma · Co-deletion of chromosomal arms 1p/19q · Machine learning

© Springer Nature Switzerland AG 2020
H. Mohy-ud-Din and S. Rathore (Eds.): RNO-AI 2019, LNCS 11991, pp. 61–69, 2020.
https://doi.org/10.1007/978-3-030-40124-5_7

1   Introduction

Gliomas are the most common tumors originating in the brain [1], and are classified into four grades, depending on their aggressiveness, by World Health Organization (WHO). Lower grade gliomas (LGG) comprising WHO grades II and III, also called diffuse low-grade and intermediate-grade gliomas, include oligodendrogliomas, astrocytomas, and oligoastrocytomas. When compared with high-grade gliomas, such as WHO grade IV glioblastoma tumors, LGG show less aggressiveness and favorable prognosis.

Different groups of genetic aberrations are associated with LGGs, and with varying sensitivity to targeted therapies [1]. A very prevalent and important genetic alteration in LGGs is the co-deletion of 1p/19q chromosome arms. Several studies [2–5] have shown that co-deletion of 1p/19q chromosome arms has prognostic value in estimating positive response of the tumor to chemotherapy and radiotherapy, both, in LGG and is correlated with favorable outcome. Therefore, predicting 1p/19q status has very important implications for adopting effective treatment regimens for LGG.

Determination of 1p/19q so far has required *ex-vivo* postoperative or biopsy tissue analyses, which are limited in assessing the tumor's spatial heterogeneity (sampling error due to single sample histopathological and molecular analysis) and temporal heterogeneity (not possible to continuously assess the molecular transformation of the tumor during treatment). Recent studies have shown some promise that 1p/19q status can be *non-invasively* detected using imaging [6–8], however mostly these studies have performed only univariate analysis.

In this study, we aim to determine the presence of 1p/19q co-deletion in a non-invasive way in patients with LGGs, grade II and III, by advanced multivariate pattern analysis of preoperative MRI scans. We employed linear kernel of support vector machines (SVMs) [9] to carry out this multivariate analysis and to develop a predictive model to determine the presence of 1p/19q mutation. This work is different from prior work in that it uses multi-parametric imaging and utilizes comprehensive imaging measures and advanced pattern analysis. It has been increasingly shown in the past that advanced pattern analysis of MRI scans, where different imaging sequences provide comprehensive and complementary information, can provide very comprehensive characterizations of gliomas [10–14]. In addition, these advanced pattern analysis methods have been proven to non-invasively detect molecular markers of glioblastoma [15, 16]. We hypothesize that quantification of comprehensive imaging features extracted from MRI scans may lead to non-invasive determination of molecular characteristics of the tumors, which is 1p/19q co-deletion status in our case, with very promising success rate on an individual patient basis.

2   Materials and Methods

2.1   Data Acquisition

Preoperative baseline mpMRI scans of 159 patients, comprising 102 1p/19q co-deleted and 57 non-co-deleted cases, were downloaded from The Cancer Imaging Archive. The data was divided into independent discovery ($n = 75$, 55 co-deleted, 30 non-co-deleted) and replication ($n = 74$, 47 co-deleted, 27 non-co-deleted) cohorts. mpMRI data included T1-weighted, and T2-weighted (T2) MRI scans. Standard of care treatment, including maximal possible safe resection, radiotherapy, chemotherapy with tomozolomide was provided to all the patients enrolled in this study.

2.2   Image Preprocessing Applied on the Dataset

All MRI of each patient were pre-processed using a series of image processing steps, including: (i) smoothing (i.e., reducing high frequency noise variations while preserving underlying anatomical structures) using Smallest Univalue Segment Assimilating Nucleus (SUSAN) denoising [17]; (ii) correction for magnetic field inhomogeneity using N3 bias correction [18]; (iii) co-registration of all MRIs of each patient at 12-degrees of freedom for examining anatomically aligned signals at the voxel level using affine registration through the Linear Image Registration Tool [19]; (iv) skull stripping using the Brain Extraction Tool [20]; and (v) matching of intensity profiles (histogram matching) of all MRIs of all patients to the corresponding MRIs of a reference patient in a linear fashion.

2.3   Segmentation of Sub-regions of Tumor

A well-established semi-automatic segmentation method based on random forest, provided by ITK-SNAP (itksnap.org) [21], was used to segment various sub-regions of the tumor, including enhancing tumor (TU), non-enhancing tumor (nTU) core, and edema region (EM). The segmentations were then registered to a standard atlas template in order to produce a standardized statistical distribution atlas for quantifying the tumor spatial location. The segmentations were assessed and revised before image analysis, when necessary.

2.4   Quantitative Imaging Features Extracted from MRI Scans

The segmentation of tumor sub-regions was followed by the extraction of relevant quantitative imaging features for each subject from all scans, to capture phenotypic imaging characteristics of the tumors and create the proposed imaging signature of 1p/19q. The features included: (i) first-order statistical moments comprising mean, median, standard deviation, skewness and kurtosis, (ii) distribution of the voxel intensities within the region and several measures derived from the intensity distributions such as maximum, minimum of the voxel intensities in all the imaging sequences, and the number of voxels in different bins of the intensity distributions, (iii) principal component analysis of histograms of various imaging signal distributions, (iv) texture features, describing the second-order statistical moments, computed from the gray-level co-occurrence matrix (GLCM) [22] and grey-level run length matrix (GLRLM) [23]. In order to calculate texture features, the intensity of all the scans was first quantized to 16 different gray-level values, and then a bounding box of 5 voxels in all the three dimensions was used as a sliding window. GLCM and GLRLM matrices were then populated by considering the intensity values within a radius of 2 pixels and for the 13 main 3D directions. Several texture features comprising energy, contrast, entropy, dissimilarity, homogeneity, and inverse difference moment, and correlation were extracted for each direction and averaged to compute the final value per region.

2.5   Spatial Distribution and Pattern of the Tumor

Two spatial frequency maps (SFMs), one for each 1p/19q status, i.e., SFM+ and SFM− for 1p/19q co-deleted and 1p/19q non-co-deleted, respectively, were constructed to capture the spatial distribution of tumors, as suggested earlier [24]. These SFMs were developed by overlaying the tumor segmentation for all patients pertaining to their 1p/19q status, and were then used to calculate the similarity of a new tumor with each of the developed atlases. Four discrete quantitative features were then derived for each new tumor, based on the average and maximum distribution of each SFM for the tumor area:

1. max(SFM + (Tumor))/max(SFM − (Tumor))
2. average(SFM + (Tumor))/average(SFM − (Tumor))
3. max(SFM + (Tumor))−max(SFM − (Tumor))
4. average(SFM + (Tumor))−mean(SFM − (Tumor))

Moreover, the percentage of tumor in each lobe of the brain, and the distance of various subregions of the tumor from ventricles have also been quantified and used as features.

2.6   Selection of Features and Development of Predictive Model

Support vector machines (SVM) was used to integrate all the imaging features via a leave-one-out (LOO) cross-validation, and to determine the combination of features most predictive of 1p/19q, separately in discovery and validation cohorts, and also in complete cohort. The LOO scheme assesses any cohort as it was divided in independent training and testing cohorts, but in a statistically robust manner. In each iteration of LOO scheme, it divides the cohort in training (n-1 samples) and testing (nth sample) subsets and then repeating the same process until all the n samples have been tested. In order to estimate the generalizability of the proposed model and to assess its behavior on unseen patient data, we trained a model on the discovery cohort and applied it on the patients of the validation cohort.

The linear kernel function of SVM classifier was trained as we did in our previous studies [25, 26]. In all the experiments, the soft margin cost function (C) of linear kernel was optimized on the training data, using grid-search in a 10-fold cross-validated manner; $C = 2\beta$, where $\beta \epsilon [-10,10]$. By adjusting this parameter, we can control the effect of each individual support vector, and can establish a tradeoff between error plenty and stability of the classifier.

3   Results and Application

3.1   Classification Performance of Predictive Model

The data was divided into discovery and validation cohorts, where discovery cohort comprised of 55 co-deleted and 30 non-co-deleted patients, and validation cohort comprised of 47 co-deleted and 27 non-deleted patients. The accuracy of our model in correctly predicting the 1p/19q mutation status on the discovery and validation cohorts was 88.24% (sensitivity = 89.09%, specificity = 86.67%) and 85.14% (sensitivity = 85.11%, specificity = 85.19%), respectively. The predictive models' performance in correctly predicting the 1p/19q status in the complete cohort was 86.16% (sensitivity = 88.24%, specificity = 82.46%).

Moreover, to determine the generalizability of the proposed model on unseen population, feature selection and model development in the discovery cohort was done by LOO cross-validation, and the performance in the validation cohort was calculated by applying the model trained in the discovery cohort in an independent fashion. The performance in this case was 82.43% (sensitivity = 82.97%, specificity = 81.48%).

Table 1. Performance of the proposed predictive model on discovery and validation cohorts, S.E. stands for Standard Error, and CI stands for confidence interval.

	Accuracy	Sensitivity	Specificity	AUC, S.E. [95% CI]
Discovery (LOO)	88.24	86.67	89.09	0.84, 0.04 [0.75–0.92]
Validation (LOO)	85.14	85.19	85.11	0.86, 0.04 [0.78–0.95]
Combined cohort (LOO)	86.16	82.46	88.24	0.86, 0.03 [0.81–0.92]
Validation (Split)	82.43	81.48	82.97	0.82, 0.04 [0.73–0.92]

The performance of the various models developed here on the discovery and validation cohorts was also assessed using receiver operating characteristic (ROC) analysis. The area under the ROC curve (AUC) was calculated for each of the models developed herein, and is shown in Table 1 and Fig. 1.

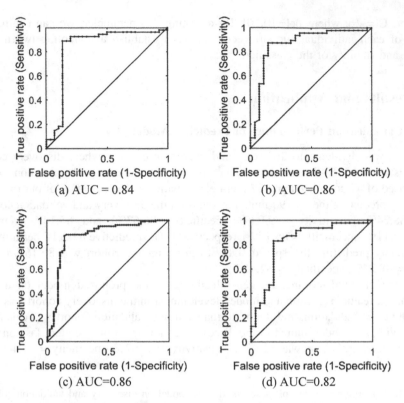

Fig. 1. ROC curves in discovery (a), replication (b) and complete (c) cohorts using LOO cross-validation. ROC curve obtained by training the model on discovery cohort and independently evaluating on the replication cohort (d).

3.2 Selected Quantitative Features

The selected features, indicated by our model in different iterations of the cross-validation, include spatial frequency distribution of the tumor in different brain regions and intensity distribution in different sub regions of the tumor. The spatial distribution of tumors of non-co-deleted and co-deleted groups is given in Fig. 2, which indicates that non-co-deleted tumors are mostly prevalent in frontal lobe of the brain, whereas co-deleted tumors show predilection to temporal lobe of the brain.

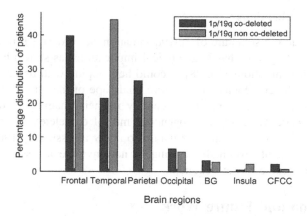

Fig. 2. Spatial frequency distribution of tumors in different regions of the brain. X-axis shows the percentage distribution, and y-axis shows the brain regions (BG = Basal ganglia, CFCC = Cingulate fornix corpus callosum).

Individual selected features accompanied by their AUC determined by ROC curve analysis and statistical significance based on a univariate analysis in the complete cohort are reported in Table 2. These features were also shown to be better predictors of various mutations in lower-grade gliomas in a previous study [8], thereby confirming these findings. These AUC and p-values can be used to determine the discriminating capability of each individual feature.

Table 2. Features discriminating between 1p/19q non-co-deleted and co-deleted tumors, quantified in terms of AUC and Effect Size method [27]. Top-most 10 features are shown here.

Feature	AUC	Effect size
Age	0.671	0.001
T1_Histogram_Mean	0.663	0.003
T1_Histogram_Homogeneity	0.660	0.008
Location_Frontal	0.641	0.008
Location_Temporal	0.632	0.009
Location_Max[+]/Max[−]	0.631	0.010
T1_GLRLM_HGRE	0.625	0.013
T1_Histogram_Percentile90	0.621	0.014
T1_GLCM_Correlation	0.620	0.017
Shape_Elongation	0.618	0.018
Shape_Perimeter	0.608	0.023
T2_Histogram_Std.Deviation	0.604	0.026
T2_Histogram_Contrast	0.600	0.029
T2_GLRLM_SGRE	0.598	0.034
T2_GLCM_Contrast	0.596	0.034

3.3   Clinical Implications

The proposed radiomic signature of 1p/19q mutation status derived from preoperative MRI of lower grade glioma has huge clinical implications, as shown by the radiomic signatures of other mutations [10, 28]. It could help coping with tissue sampling error by evaluating the heterogeneity in the entire landscape of the tumor. It can also help identifying tumors having higher heterogeneity in their imaging feature, eventually pinpointing the patients having heterogeneous mix of co-deleted and non-co-deleted tissue. Finally, the proposed signature may also allow assessing 1p/19q status before and after the treatment, eventually leading to non-invasive assessment of mutation status during the course of the treatment.

4   Conclusion and Future Work

We utilized advanced pattern analysis on preoperative MRI scans of glioma patients to derive non-invasive imaging signature of 1p/19q mutation status. Our results are very promising and have the potential to cope with the limitations of tissue sampling errors and guide patient management at the initial presentation of the disease. We did not utilize advanced imaging, Dynamic Susceptibility Contrast MRI and Diffusion Tensor Imaging, which shows that our proposed radiomic signature can predict the 1p/19q mutation status using conventional/basic imaging only. We anticipate that the synergistic use of conventional and advanced imaging in future may further improve the prediction rates. The current study has been conducted on a small population of n = 159 patients, however, we believe that larger datasets might lead to development of a robust predictive model and better prediction of 1p/19q status. Moreover, although our proposed radiomic signature successfully predicted 1p/19q mutation status, the individual features selected by the model need to be evaluated and their biological significance need to be explored.

References

1. Louis, D., et al.: World Health Organization classification of tumours of the central nervous system. In: International Agency for Research on Cancer, Lyon, p. 4 (2007)
2. Fellah, S., et al.: Multimodal MR imaging (diffusion, perfusion, and spectroscopy): is it possible to distinguish oligodendroglial tumor grade and 1p/19q codeletion in the pretherapeutic diagnosis? AJNR Am. J. Neuroradiol. **34**, 1326–1333 (2013)
3. Jansen, N.L., et al.: Prediction of oligodendroglial histology and LOH 1p/19q using dynamic [(18)F]FET-PET imaging in intracranial WHO grade II and III gliomas. Neuro Oncol. **14**, 1473–1480 (2012)
4. Iwadate, Y., et al.: Molecular imaging of 1p/19q deletion in oligodendroglial tumours with 11C-methionine positron emission tomography. J. Neurol. Neurosurg. Psychiatry **87**, 1016–1021 (2016)
5. Bourdillon, P., et al.: Prediction of anaplastic transformation in low-grade oligodendrogliomas based on magnetic resonance spectroscopy and 1p/19q codeletion status. J. Neurooncol. **122**, 529–537 (2015)
6. Akkus, Z., et al.: Predicting deletion of chromosomal arms 1p/19q in low-grade gliomas from MR images using machine intelligence. J. Digit. Imaging **30**, 469–476 (2017)

7. Zhou, H., et al.: MRI features predict survival and molecular markers in diffuse lower-grade gliomas. Neuro Oncol. **19**, 862–870 (2017)
8. Chaddad, A., et al.: Predicting the gene status and survival outcome of lower grade glioma patients with multimodal MRI features. IEEE Access **7**, 75976–75984 (2019)
9. Chang, C.C., Lin, C.J.: LIBSVM: a library for support vector machines. ACM Trans. Intell. Syst. Technol. (TIST) **2**, 27 (2011)
10. Rathore, S., et al.: Radiomic MRI signature reveals three distinct subtypes of glioblastoma with different clinical and molecular characteristics, offering prognostic value beyond IDH1. Sci. Rep. **8**, 5087 (2018)
11. Rathore, S., et al.: A radiomic signature of infiltration in peritumoral edema predicts subsequent recurrence in glioblastoma: implications for personalized radiotherapy planning. J. Med. Imaging **5**, 021219 (2018)
12. Macyszyn, L., et al.: Imaging patterns predict patient survival and molecular subtype in glioblastoma via machine learning techniques. Neuro Oncol. **18**, 417–425 (2016)
13. Rathore, S., et al.: Technical note: a radiomic signature of infiltration in peritumoral edema predicts subsequent recurrence in glioblastoma. In: Medical Imaging 2018: Image-Guided Procedures, Robotic Interventions, and Modeling, vol. 10576, p. 105760O (2018)
14. Shukla-Dave, A., et al.: The utility of magnetic resonance imaging and spectroscopy for predicting insignificant prostate cancer: an initial analysis. BJU Int. **99**, 786–793 (2007)
15. Rathore, S., et al.: Multivariate pattern analysis of de novo glioblastoma patients offers in vivo evaluation of O6-methylguanine-DNA-methyltransferase (MGMT) promoter methylation status, compensating for insufficient specimen and assay failures. J. Neuro-oncol. **20**, vi186 (2018)
16. Bakas, S., et al.: In vivo detection of EGFRvIII in glioblastoma via perfusion magnetic resonance imaging signature consistent with deep peritumoral infiltration: the phi-index. Clin. Cancer Res.: Off. J. Am. Assoc. Cancer Res. **23**, 4724–4734 (2017)
17. Smith, S.M., Brady, J.M.: SUSAN - a new approach to low level image processing. Int. J. Comput. Vis. **23**, 45–78 (1997)
18. Tustison, N.J., et al.: N4ITK: improved N3 bias correction. IEEE Trans. Med. Imaging **29**, 1310–1320 (2010)
19. Jenkinson, M., Smith, S.: A global optimisation method for robust affine registration of brain images. Med. Image Anal. **5**, 143–156 (2001)
20. Smith, S.M.: Fast robust automated brain extraction. Hum. Brain Mapp. **17**, 143–155 (2002)
21. Yushkevich, P.A., et al.: User-guided 3D active contour segmentation of anatomical structures: significantly improved efficiency and reliability. NeuroImage **31**, 1116–1128 (2006)
22. Haralick, R.M., et al.: Textural features for image classification. IEEE Trans. Syst. Man Cybern. **3**, 610–621 (1973)
23. Galloway, M.M.: Texture analysis using grey level run lengths. Comput. Graph. Image Process. **4**, 172–179 (1975)
24. Bilello, M., et al.: Population-based MRI atlases of spatial distribution are specific to patient and tumor characteristics in glioblastoma. Neuroimage Clin. **12**, 34–40 (2016)
25. Rathore, S., et al.: GECC: gene expression based ensemble classification of colon samples. IEEE/ACM Trans. Comput. Biol. Bioinf. **11**, 1131–1145 (2014)
26. Rathore, S., et al.: Automated colon cancer detection using hybrid of novel geometric features and some traditional features. Comput. Biol. Med. **65**, 279–296 (2015)
27. Sullivan, G.M., Feinn, R.: Using effect size-or why the P value is not enough. J. Grad. Med. Educ. **4**, 279–282 (2012)
28. Verhaak, R.G., et al.: Integrated genomic analysis identifies clinically relevant subtypes of glioblastoma characterized by abnormalities in PDGFRA, IDH1, EGFR, and NF1. Cancer Cell **17**, 98–110 (2010)

A Feature-Pooling and Signature-Pooling Method for Feature Selection for Quantitative Image Analysis: Application to a Radiomics Model for Survival in Glioma

Zhenwei Shi[1]([✉]), Chong Zhang[1], Inge Compter[1], Maikel Verduin[2],
Ann Hoeben[2], Danielle Eekers[1], Andre Dekker[1], and Leonard Wee[1]

[1] Department of Radiation Oncology (MAASTRO),
GROW School for Oncology and Developmental Biology,
Maastricht University Medical Centre+, Maastricht, The Netherlands
`zhenwei.shi@maastro.nl`
[2] Department of Medical Oncology,
GROW School for Oncology and Developmental Biology,
Maastricht University Medical Centre+, Maastricht, The Netherlands

Abstract. We proposed a pooling-based radiomics feature selection method and showed how it would be applied to the clinical question of predicting one-year survival in 130 patients treated for glioma by radiotherapy. The method combines filter, wrapper and embedded selection in a comprehensive process to identify useful features and build them into a potentially predictive signature. The results showed that non-invasive CT radiomics were able to moderately predict overall survival and predict WHO tumour grade. This study reveals an associative inter-relationship between WHO tumour grade, CT-based radiomics and survival, that could be clinically relevant.

Keywords: Quantitative imaging feature · Feature selection · Glioma

1 Introduction

With ever-increasing utilization of radiological imaging in the diagnosis and treatment workflow for deadly cancers such as glioblastoma, there is intense interest in use of quantitative image analysis to extract as much clinically relevant information as possible for research and routine care. Despite the growing repertoire of drug agents and radiotherapy tools available to oncologists for treating non-resectable brain tumours, survival among glioblastoma patients remains

Electronic supplementary material The online version of this chapter (https:// doi.org/10.1007/978-3-030-40124-5_8) contains supplementary material, which is available to authorized users.

depressingly low, with median survival time just over 14 months and as little as 10% of patients still alive after 5 years.

Radiomics [1,4] is an emerging field of translational clinical research aiming to convert vast amounts of routine clinical imaging data into a mineable "big data" resource to promote research into better treatments and to derive actionable insights to guide medical decision-making. Image-derived metric, i.e. radiomics features, are though to encode information about cancer and its phenotype, using subtle characteristics of pictures that are not easily quantifiable by an unaided human eye. Among imaging modalities, computed tomography (CT) radiomics is widely used, due to the relatively universal coverage of CT scanners in oncology centres [13]. However, reproducibility of radiomics studies across multiple independent institutions remains a significant translational research challenge, though good reproducibility is generally reported for CT [13] and rapid progress is being made in the magnetic resonance imaging (MRI) [9] and positron emission imaging (PET) radiomics domain [12].

Generally, radiomics analysis relies heavily on supervised machine learning, and the process can be divided into a number of essential parts: (i) image acquisition (ii) identification of the tumour region of interest (iii) radiomics feature extraction, (iv) selection of potentially predictive features from a vast set of computed features, and finally (v) model development and validation [6].

Among these steps, robust and comprehensive methods for feature selection plays a major role in regards to reducing risk of over-fitting and developing a compact, parsimonious final model that only contains the essential predictive variables. Feature selection in machine learning generally comes in three distinct flavours of method, each with their own advantages and disadvantages – filter methods (e.g. chi-squared test), wrapper methods (e.g. stepwise feature inclusion/elimination) and embedded methods (e.g. regularization) [10].

Previous work in this area have compared combinations of feature selection approaches and machine classifiers. Hawkins et al. compared four different feature selection and classification methods for CT-based survival prediction of lung cancer [5]. Parmar et al. first evaluated 14 feature selection methods and 12 classification methods to predict overall survival of lung cancer patients [7], then chose 13 feature selection methods and 11 classification methods to predict overall survival of head and neck cancer patients [8]. Zhang et al. compared 54 cross-combinations of six feature selection methods and nine classification methods for prediction of local and distant failure in advanced nasopharyngeal carcinoma [16]. Wu et al. compared 24 feature selection and three classification methods for the prediction of lung cancer histology [15].

It is clear that inappropriate feature selection method can adversely affect the performance of a radiomics signature. However, there is no known a prior method for selecting the most predictive features from the beginning. The motivations for developing a good feature selection procedure are clear to obtain an unbiased and generalizable model. We need feature selection to deal with the two-fold problem that is common in many clinical transnational radiomics studies – a limited patient sample size (that is, a small number of outcome events) relative to the feature dimensional space which could be one or more orders of magnitude larger than the patient sample size.

In this manuscript, we propose a comprehensive feature selection procedure combining several advantages of different filter, wrapper, and embedded approaches. The intention was to develop a feature selection method that could work with different combinations of feature selection and classifier training approaches. The proposed method uses feature pooling to make transparent which individual features are frequently selected, and allows a set of competing candidate signatures to be assessed in the training set. As proof of concept, we used this method to select a CT radiomics-based signature for overall survival in glioma/glioblastoma patients. We evaluated the predictive performance of this model on a single-institution dataset, and further considered the relationships between the selected signature, tumour aggressiveness grade and overall survival at one year post-radiotherapy.

2 Materials and Methods

This was an internal ethics board-approved retrospective study comprising 160 patients with pathologically-confirmed glioma treated at a single radiotherapy institution between January 2004 and December 2014. Of these, 130 DICOM-RT studies with the requisite radiotherapy dose planning CT scans were extracted from a research picture archival system (PACS). The case mix examined here consists of 93 glioblastoma (GBM) and 37 non-GBM astrocytomas of various subtypes. All patients received only biopsy prior to high-dose radiotherapy with temozolomide or radiotherapy only. Follow-up consisted of quarterly clinical consultations including MRI examination, until death from any cause. MRI and CT images were co-registered and a gross tumour volume (GTV) was manually drawn by an experienced radiation oncologist. Radiotherapy dose and radiomics features were calculated on helical CT (Siemens, Erlangen, Germany) with 0.98 mm by 0.98 mm pixels, 1 mm reconstructed slice thickness and 120 kVp tube potential.

2.1 Feature Extraction

Image features were extracted from hand-drawn GTVs using an open-source package O-RAW [11] that is an extension wrapper for the freely available radiomics software PyRadiomics [14]. Each CT slices was uniformly resampled to isotropic voxels of 1 mm by using linear interpolation. A total of 1105 radiomic features, consisting of first-order statistics, shape descriptors, texture features, and filter-based features by Wavelet and Laplacian of Gaussian (LoG), were extracted from each 3D GTV. Documentation of pyradiomics features can be found (https://pyradiomics.readthedocs.io).

2.2   Feature Selection

We proposed an integrated feature selection procedure to minimize the risk of over-fitting the radiomics model on the available patient outcome data, and this is shown schematically in Fig. 1. The principal idea is based on pooling the features and signatures over numerous bootstrap sampling iterations. The feature pooling part was used to individually rank potentially predictive radiomics features by its cumulative selection frequency. The signature pooling part reveals the set of potentially predictive signatures built using these frequently selected individual features, rather than just one signature. For this proof of concept, we used 500 bootstrap iterations as example, and we then selected the first signature from its pool to assess the area under a receiver-operator curve (AUC) and classification accuracy. We elaborate the process in detail below.

First, the available patient cohort was randomly split into a training (75%) subset for feature selection and model development, and a validation subset (25%) for model assessment only. By counting outcome labels only, we confirmed that the ratio of surviving to deceased subjects in each subset was the same. Secondly, dimensionality reduction was performed in the training set only by testing (i) feature Pearson correlation coefficient [2] against volume (ii) pair-wise feature Pearson correlation coefficient to each other, and (iii) Kolmogorov-Smirnov (KS) non-parametric similarity of features among the surviving and deceased outcome labels. For this study, we selected a p-value of 0.05 as the elimination threshold. Thirdly, the least absolute shrinkage and selection operator (LASSO) was used as an embedded regularization and feature selection method. Within the LASSO operation, we applied the Synthetic Minority Oversampling Technique (SMOTE) [3] 500 times to ameliorate the effects of the unbalanced alive to dead ratio in the training subset.

With individual feature pooling, we kept count of the number of times each radiomic feature was retained after the previous LASSO step. We then ordered these retained features from most-frequent to least-frequent and kept only those individual features which occurred more often than the mean frequency. These relatively frequently-retained features were then subjected to 5-fold, 500-repetition recursive feature elimination (RFE) with a logistic regression classifier to compile a set of candidate radiomics signatures. Each signature thus consists of a maximally compact set of individual frequently-appearing features from the feature pool that collectively contributes to the outcome prediction. However, the RFE step above often selects the same subset of individual features. Again, we tracked the number of times signature with the same combination of features was selected at the end of RFE. We ordered these candidate signatures from most-frequently to least-frequently appearing. For this study, we arbitrarily retained the top 5 most-commonly occurring signatures for inspection. We computed the AUC of all top 5 signatures on all the subjects in the training subset only, and then chose one signature with the highest AUC for testing in the validation subset. Throughout the feature selection with LASSO and signature development with RFE, we used a multi-variable logistic regression statistical model for the binary surviving vs dead outcome prediction.

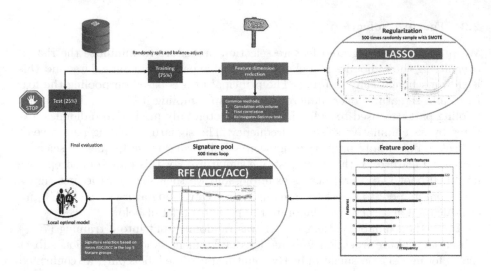

Fig. 1. The processing workflow of the proposed feature-pooling and signature-pooling method for feature selection for quantitative image analysis. The details are explained in the text.

2.3　Statistical Analysis

The above methodology was implemented in its entirety in R (version 3.30, https://www.r-project.org). 95% confidence intervals around the mean AUC and mean classification accuracy were estimated using 2000 stratified bootstrap replications. To assess goodness of fit, we assessed the slope and intercept of the final model calibration curve on the validation subset, accompanied by a Hosmer-Lemeshow test statistic.

2.4　Model Development

A prognostic model of overall survival at one year following radiotherapy treatment of glioma, using radiomics features only, was developed using the above-mentioned method. For comparison, we evaluated the AUC and accuracy of one-year survival prediction using only the World Health Organization (WHO) tumour aggressiveness grades at diagnostic baseline and at the start of radiotherapy. For comprehensiveness of our understanding of the role of radiomics, we also investigated the prediction of WHO tumour grade using radiomics features alone, following the same radiomics signature as we developed for survival.

3　Results

At one year following radiotherapy, there was similar proportion of surviving to deceased patients on the training and validation subsets (training: 64 deceased,

33 alive; training: 20 deceased, 13 alive), see Supplementary Fig. 1. After executing our feature selection process for survival prediction, we arrived at the individual radiomics features at the top of the feature pool, as shown in Supplementary Fig. 2. From the AUC metrics of the top 5 most-frequently appearing signatures in the signature pool, we selected the top AUC signature to test against the validation subset. We gave this tentative signature the label "shi:2019a-gbm" for reference.

The "shi:2019a-gbm" signature validated well, achieving an AUC of 0.82 (95% CI: 0.67–0.97) and accuracy of 0.70 (95% CI: 0.51–0.84) in the validation subset. For comparison, the same signature achieved an AUC of 0.77 (95% CI: 0.66–0.87) and accuracy of 0.73 (95% CI: 0.63–0.82) in the training subset. The difference in AUCs and classification accuracies between the training and validation subsets were not statistically significant, as can be expected from the highly-overlapping confidence interval estimates.

Figure 2 gives the AUC plots of "shi:2019a-gbm" in the training subset (black line) and in the validation subset (red line). The calibration plots of the signature in both training and validation subsets with the Hosmer-Lemeshow test are shown in Supplementary Fig. 3. The Hosmer-Lemeshow test of the one-year survival prediction model yielded non-significant statistics ($p = 0.3$ and $p = 0.12$), indicating that deviation of model prediction from observed outcome was not statistically significant. The logistic regression coefficients of the "shi:2019a-gbm" signature is shown in Fig. 3.

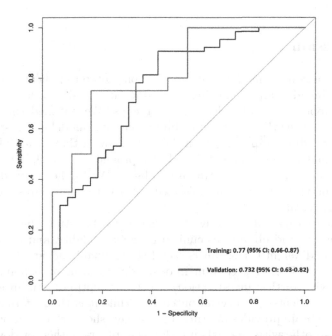

Fig. 2. Receiver-operating characteristic curves of the developed CT radiomics signature shi:2019a-gbm in the training and validation datasets. (Color figure online)

Feature name	Coefficients	Intercept = -1.784
original_glcm_Correlation	4.959	
wavelet.LHH_glszm_LowGrayLevelZoneEmphasis	-1.484	
wavelet.LHH_glrlm_LongRunHighGrayLevelEmphasis	-8.703 e^{-4}	
wavelet.LLH_gldm_SmallDependenceHighGrayLevelEmphasis	5.545 e^{-3}	
wavelet.HLL_glcm_ClusterProminence	0.65	

Fig. 3. The logistic regression coefficient of each radiomic feature in the shi:2019a-gbm sinecure with intercept.

To set a comparative baseline for the prediction performance of signature "shi:2019a-gbm", we examined the survival prediction by WHO tumour grades at diagnostic baseline and at the start of radiotherapy using the same patient data, respectively. The tumour grades yielded similar results to the radiomics signature with accuracy of 0.79 (95% CI: 0.61–0.91) and 0.70 (95% CI: 0.51–0.84) in the validation subset, respectively. When we re-calculated regression coefficients in "shi:2019a-gbm" (i.e. the same set of potentially predictive radiomics features), we found this subset of radiomic features also performs well for predicting WHO tumour grades with AUCs of 0.80 (95% CI: 0.64–0.95) and 0.73 (95% CI: 0.64–0.93) for the tumour grades at diagnostic baseline and at the start of radiotherapy.

4   Discussion

We proposed a comprehensive feature selection method applicable generally to quantitative imaging analysis using derived features, using a new combination of filter, wrapper and embedded selection steps. We outlined the method in detail, and then used this to develop a binary outcome model for one-year overall survival after radiotherapy of glioma. We thus showed that a carefully selected radiomics signature could be developed as a prognostic model, with equivalent performance in validation and training subsets. We further probed the interrelationship between the radiomics signature, WHO tumour aggressiveness grade and overall survival.

We elected to use retrospective CT images in this study, because CT is a relatively routine and ubiquitous clinical imaging modality, and reproducibility between many different CT scanners could be reasonably achieved. Our notable finding is that CT images contain information, in form of radiomic features, that could (with further investigation) provide an entirely non-invasive means of stratifying patients for survival outcome. While it is thought that only MR imaging of the brain provides information, we have shown that routine CT imaging that would otherwise be single-use (for example, radiotherapy dose planning only) can be easily re-used to extract clinically relevant information.

The inter-relationship of radiomics with WHO tumour grade is interesting, quite apart and separately from survival prediction. With only re-calibration of the coefficients, we achieved promising results in predicting WHO tumour grade re-using the same radiomic features in the signature ("shi:2019a-gbm") as we used to predict survival. This suggests a hypothesis, that the signature we selected happens to associate closely with WHO tumour grade, therefore either the tumour grade or the radiomics signature provides an similar pathway towards predicting survival at one year, at least with a large overlap. It is highly likely, if we optimized the feature selection to predict WHO tumour grade from the very beginning (instead of survival) we would be able to additionally improve the prediction of tumour grade from a fully non-invasive radiomics signature alone. However, this was not the specific objective of this paper.

Before we can suggest clinical applicability and wider generalizability of the above model, we need to perform external validation across multiple clinical centres, and further examine the inter-relationship between radiomics signatures with clinical prognostic factors. In terms of tumour volume-confounding of radiomics features, we confirmed that the individual features incorporated into the signature "shi:2019a-gbm" had effectively zero Pearson correlation (median 0.02, range: -0.27–0.18) with GTV volume.

Even though a full explanation of radiomic features for survival of glioma patients is difficult, we have tried to show the definition and biological meaning of the selected features below. These features may different from the previously studies. However, as many radiomic features are correlated with each other, we may have similar findings to others. To verify this hypothesis, further studies are still needed.

1. Original_glcm_Correlation feature is with the value between 0 (uncorrelated) and 1 (perfectly correlated) showing the linear dependency of gray level values to their respective voxels in the GLCM. Therefore, it may reflect the density and homogeneity of the tumour.
2. Wavelet.LHH_glszm_LowGrayLevelZoneEmphasis measures the distribution of lower gray-level size zones, with a higher value indicating a greater proportion of lower gray-level values and size zones in the image. Therefore, it may reflect the homogeneity of the tumour.
3. Wavelet.LHH_glrlm_LongRunHighGrayLevelEmphasis measures the joint distribution of long run lengths with higher gray-level values. Therefore, it may reflect the homogeneity of the tumour as well.
4. Wavelet.LLH_gldm_SmallDependenceHighGrayLevelEmphasis measures the joint distribution of small dependence with higher gray-level values. Unfortunately, we do not come up with the biological meaning to the tumour.
5. Wavelet.HHL_glcm_ClusterProminence is a measure of the skewness and asymmetry of the GLCM. A higher values implies more asymmetry about the mean while a lower value indicates a peak near the mean value and less variation about the mean. It may reflect the asymmetry and homogeneity of the tumour.

To increase the chances of clinical applicability and wider generalizability of radiomics, appropriate methods for feature selection and signature compilation are much needed in the field. Previous workers [5, 7, 8, 15, 16] have investigated the consequences of feature selection on radiomics model performance. We have tried to overcome some limitation of previous studies using our proposed approach. First, restricting oneself to a simple feature selection step might not be able to locate relevant signals associated with the outcome. Secondly, being overly profligate with feature dimensionality might lead to over-fitting, which will be observed when testing a signature in hold-out or external validation data. Thirdly, our feature-pooling and signature-pooling steps allow information derived from each bootstrap sample to be used to rank the most promising features. Pooling makes it transparent to the investigator that certain combinations of features are more likely to be encountered than others. This makes the feature selection method relatively independent of "lucky" happenstances in the way the cohort data is randomly divided.

It is possible that different learning algorithms may work better with certain feature selection methods. For example, selecting feature by a regularized linear regression approach may not be appropriate if one is ultimately intending to develop a non-linear classifier. Therefore, some consistency may be required between the feature selection steps and the model learning steps. Additionally, there are potential limitations when using penalty functions in regularization methods with high-dimensionality on data with limited number of outcomes. For instance, when the dimension P of radiomics is much larger than N the number of patients, a LASSO method can choose at most N features before it saturates. In this study, totally 1105 radiomic features were extracted, but only 97 patients were in the training dataset. Hence, the feature dimension reduction step was extremely necessary, which made LASSO work correctly. Furthermore, if there is a group of highly correlated features, then the LASSO method tends to select one feature from a group and ignore the others. To overcome this, we tracked the cumulative frequency of retention of individual features first, rather than using LASSO to select the final feature set for us.

In our knowledge, there is no a baseline approach for radiomic feature selection, as the feature selection procedure is highly dependent on radiomics feature dimension (e.g., hundreds to thousands), sample size and different categories of radiomics. This study described a new feature selection framework, especially proposing feature pooling and signature pooling approaches. These two steps are reasonable and feasible to select robust radiomic features so that avoiding over-fitting. For the algorithms in the framework, such as KS test, LASSO, and REF, were suitable and worked well in our study. We showed them as enlightenment to new users, which could help them to start to use the proposed framework. Certainly, users could replace them with other relevant algorithms according to their specific cases.

The key limitation of this study is the lack of a fully independent external cohort from another radiotherapy institution. To test the methodology, we have only used hold-out validation on the single-institution cohort. For future work, we are preparing additional validation cohorts from other institutions in order to externally validate and challenge our hypotheses. Testing of this method on other diseases and alternative clinical endpoints was outside the scope of the present study, however we are preparing lung cancer and oropharyngeal cancer data for several hundred subjects that we will share publicly. This feature selection method can be tested on those datasets in the near future.

A detailed comparison of different classifiers matched to different options for feature selection was outside the scope of this paper. Furthermore, combination of the radiomics signature with clinical variables (e.g., WHO tumour grade) still needs to be considered for future work. However, the method we have proposed here is modular in the sense that feature selection steps and model training steps can be changed without affecting the overall process. Thus, the pooling functionality can then be used to transparently document whether alternative choices of feature selection might produce different signatures.

5   Conclusion

We have proposed a pooling-based radiomics feature selection method, and show how it would be applied to the clinical question of predicting 1-year survival in patients treated for glioma by radiotherapy. The method combines filter, wrapper and embedded selection in a comprehensive process to identify useful features and build them into a potentially predictive signature. Non-invasive CT radiomics features were able to moderately predict overall survival and predict WHO tumour grade. Subject to further validation in future, this study reveals an associative inter-relationship between WHO tumour grade, CT-based radiomics and survival, that could be clinically relevant.

References

1. Aerts, H.J., et al.: Decoding tumour phenotype by noninvasive imaging using a quantitative radiomics approach. Nat. Commun. **5**, 4006 (2014)
2. Benesty, J., Chen, J., Huang, Y., Cohen, I.: Pearson correlation coefficient. In: Cohen, I., Huang, Y., Chen, J., Benesty, J. (eds.) Noise Reduction in Speech Processing. STSP, vol. 2, pp. 1–4. Springer, Berlin (2009). https://doi.org/10.1007/978-3-642-00296-0_5
3. Chawla, N.V., Bowyer, K.W., Hall, L.O., Kegelmeyer, W.P.: Smote: synthetic minority over-sampling technique. J. Artif. Intell. Res. **16**, 321–357 (2002)
4. Gillies, R.J., Kinahan, P.E., Hricak, H.: Radiomics: images are more than pictures, they are data. Radiology **278**(2), 563–577 (2015)
5. Hawkins, S.H., et al.: Predicting outcomes of nonsmall cell lung cancer using CT image features. IEEE Access **2**, 1418–1426 (2014)
6. Lambin, P., et al.: Radiomics: the bridge between medical imaging and personalized medicine. Nat. Rev. Clin. Oncol. **14**(12), 749 (2017)
7. Parmar, C., Grossmann, P., Bussink, J., Lambin, P., Aerts, H.J.: Machine learning methods for quantitative radiomic biomarkers. Sci. Rep. **5**, 13087 (2015)

8. Parmar, C., Grossmann, P., Rietveld, D., Rietbergen, M.M., Lambin, P., Aerts, H.J.: Radiomic machine-learning classifiers for prognostic biomarkers of head and neck cancer. Front. Oncol. **5**, 272 (2015)

9. Peerlings, J., et al.: Stability of radiomics features in apparent diffusion coefficient maps from a multi-centre test-retest trial. Sci. Rep. **9**(1), 4800 (2019)

10. Saeys, Y., Inza, I., Larrañaga, P.: A review of feature selection techniques in bioinformatics. Bioinformatics **23**(19), 2507–2517 (2007)

11. Shi, Z.: O-RAW: ontology-guided radiomics analysis workflow (2017). https://gitlab.com/UM-CDS/o-raw

12. Traverso, A., et al.: Stability of radiomic features of apparent diffusion coefficient (ADC) maps for locally advanced rectal cancer in response to image pre-processing. Phys. Med. **61**, 44–51 (2019)

13. Traverso, A., Wee, L., Dekker, A., Gillies, R.: Repeatability and reproducibility of radiomic features: a systematic review. Int. J. Radiat. Oncol. *Biol.* Phys. **102**(4), 1143–1158 (2018)

14. van Griethuysen, J.J., et al.: Computational radiomics system to decode the radiographic phenotype. Cancer Res. **77**(21), e104–e107 (2017)

15. Wu, W., et al.: Exploratory study to identify radiomics classifiers for lung cancer histology. Front. Oncol. **6**, 71 (2016)

16. Zhang, B., et al.: Radiomic machine-learning classifiers for prognostic biomarkers of advanced nasopharyngeal carcinoma. Cancer Lett. **403**, 21–27 (2017)

Radiomics-Enhanced Multi-task Neural Network for Non-invasive Glioma Subtyping and Segmentation

Zhiyuan Xue[iD], Bowen Xin[iD], Dingqian Wang[iD], and Xiuying Wang$^{(\boxtimes)}$[iD]

School of Computer Science, The University of Sydney, Sydney, Australia
`xiu.wang@sydney.edu.au`

Abstract. Non-invasive glioma subtyping can provide diagnostic support for pre-operative treatments. Traditional radiomics method for subtyping is based on hand-crafted features, so the capacity of capturing comprehensive features from MR images is still limited compared with deep learning method. In this work, we propose a radiomics enhanced multi-task neural network, which utilizes both deep features and radiomic features, to simultaneously perform glioma subtyping, and multi-region segmentation. Our network is composed of three branches, namely shared CNN encoder, segmentation decoder, and subtyping branch, constructed based on 3D U-Net. Enhanced with radiomic features, the network achieved 96.77% for two-class grading and 93.55% for three-class subtyping over the validation set of 31 cases, showing the potential in non-invasive glioma diagnosis, and achieved better segmentation performance than single-task network.

Keywords: Glioma · Radiomics · Multi-task neural network · Non-invasive subtyping · Segmentation

1 Introduction

Gliomas are the most common type of primary brain tumors. These tumors are heterogeneous in nature, having different biological characteristics, which can be characterized into subtypes based on mutations in the TERT promoter, mutations in IDH, and 1p/19q codeletion [9]. It is of great significance to accurately subtype the gliomas because these subtypes of gliomas are shown to have different responses to radiotherapy and chemotherapy, thus influencing the overall survival. Currently, accurate grading and subtyping of gliomas can only be obtained by pathological examinations, which often happens after surgery. Precise grading and subtyping predictions based on non-invasive examinations will provide diagnostic support for pre-operative treatments.

As radiomics is known for its ability to quantify the image features by applying a large number of image characterization algorithms on the region of interest (ROI) of the image, it has been applied to grade and subtype the gliomas and

H. Mohy-ud-Din and S. Rathore (Eds.): RNO-AI 2019, LNCS 11991, pp. 81–90, 2020.
https://doi.org/10.1007/978-3-030-40124-5_9

achieved promising results [6,20,22–24]. However, there are still some limitations. On the one hand, some of the radiomics works are based on the ROI manually delineated, which is a laborious task for radiologists. On the other hand, all features are extracted by hand-crafted algorithms, which might not capture all useful characteristics.

Recently, convolutional neural network (CNN) based methods [14,19] achieved more accurate segmentation results than traditional computer vision methods, and some of these methods have successfully been used to extract deep features for further analysis [12,13]. But there is a problem with this sequential approach, because the CNN is trained only to perform segmentation, the features it extracts are not optimized for other tasks such as grading or subtyping.

In this work, we propose a multi-task neural network, built based on 3D U-Net [7], a successful segmentation CNN architecture, that extracts deep features, and when enhanced with radiomic features, improves the accuracy of glioma subtyping, as well as generates multi-region segmentation for other analytical tasks. Different from the approach of using a pre-trained segmentation CNN to extract features, our network was trained to do both segmentation and classification at the same time.

2   Related Work

2.1   Grading and Subtyping Gliomas Using Radiomic Methods

Radiomics is to use mathematical methods to extract quantitative phenotypic data from medical images [1,10], and radiomic methods for grading and subtyping gliomas are to categorize gliomas into different grades and subtypes using radiomic features. Recent studies showed promising results on the grading of Lower Grade Glioma (LGG) and Higher Grade Glioma/Glioblastoma Multiforme (HGG/GBM) [6,20,24], and identifying isocitrate dehydrogenase (IDH) mutation in LGG or HGG [22,23]. These traditional radiomics methods achieved grading accuracy 89.81% [6], and subtyping accuracy 80.0% based on IDH mutation [22]. Our previous work [21] shows good results in radiomics-based pathology grading using random forest (RF)-based feature selection method combined with backward feature elimination, so in this work we used the same approach as a baseline.

Seeing the good performance of segmentation CNNs, some studies [12,13] started to use them to extract features for other analytical tasks in conjunction with hand-crafted radiomic features, and achieved better results in both diagnosis (IDH-1 mutation prediction) and prognosis (survival prediction). These studies show that there are more phenotypic features captured by segmentation CNNs but not by hand-crafted radiomic methods, and these features are also strongly correlated with the underlying nature of the tumor.

2.2   Automated Brain Tumor Segmentation

As an initial step of computer-aided diagnosis of brain tumors, automated segmentation has drawn a lot of attention, and many good methods have been proposed. Recently, CNN based methods (Fully convolutional networks [14] and U-Net [19]) have become the new state-of-the-art methods in image segmentation tasks. In Multimodal Brain Tumor Segmentation Challenges (BraTS) [4,5,16] 2018, we can see that the best segmentation results [11,15,18,25] were achieved by variations of 3D U-Net. In particular, the combined architecture of U-Net and variational autoencoder, which won the BraTS 2018 segmentation challenge [18], shows that an additional output branch (a decoder branch in this case) can help the feature extraction part (encoder part) of the network to catch more important features, thus improves the segmentation result.

Previous methods of MRI-based glioma grading/subtyping rely on either manual segmentation, which is a laborious task, or pre-trained segmentation neural network, which may introduce additional errors. We aim to improve the latter approach by combining the segmentation process and the grading/subtyping process into the same neural network, so that each process will also perform as an additional constraint to the other, thus reducing the error rate for both tasks.

3 Methods

Our radiomics enhanced multi-task neural network is made of three components: shared CNN encoder, segmentation decoder and subtyping branch, as shown in Fig. 1. The network takes four-modality MRIs as input to the shared CNN encoder, and optionally radiomic features as enhancement input to the subtyping branch, and produces segmentation results from segmentation decoder and subtyping results from subtyping branch simultaneously.

3.1 Radiomic Feature Extraction

Based on three Volume of Interest (VOI) in manual segmentation, a set of radiomic features were computed from each of four MR images, for every patient using an automatic pipeline. Every set of radiomic features contained 1124 features, including 18 first-order features, 61 texture features, 13 shape features, 688 wavelet features and 344 Laplacian of Gaussian (LoG) features. First-order and texture features are descriptive statistics based on the gray level intensity distribution of the MR images. Shape features are derived from the triangle mesh generated from the VOI, e.g. surface area, sphericity. Wavelet and LoG features are descriptive statistics calculated from images filtered by wavelet filter and LoG filter.

3.2 Shared CNN Encoder and Segmentation Decoder

The shared CNN encoder (a) and segmentation decoder (b) in Fig. 1 together, form the base structure of a 3D U-Net [7]. The shared CNN encoder has eight 3 × 3 × 3 3D convolutional layers and three 3D max-pooling layers. The segmentation decoder starts from the last layer of the encoder, following by three decoding

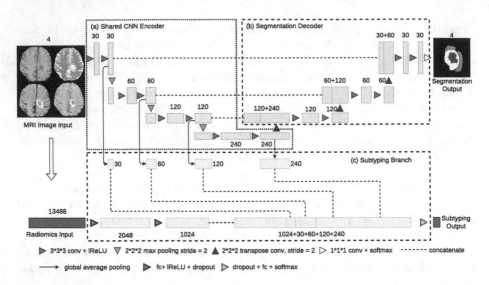

Fig. 1. The overall architecture of our multi-task neural network. It can take multi-modality MRIs together with radiomic features as input, and output segmentation and subtyping results simultaneously.

blocks, each of which has a 3D transpose convolutional layer, concatenation with corresponding convolutional layer output from the encoder, and two $3 \times 3 \times 3$ 3D convolutional layers, and finally one $1 \times 1 \times 1$ 3D convolutional layer with softmax activation function to give segmentation results. Following [11], we used leaky ReLU with 0.01 leakiness rate as activation function, and 10^{-5} weight decay as regularization. As for the loss function of the segmentation decoder, we use Dice loss (Eq. 1), which was firstly proposed for segmentation tasks in [17].

$$\mathcal{L}_{dc} = -\frac{2}{|K|} \sum_{k \in K} \frac{\sum_i u_i^k v_i^k}{\sum_i u_i^k + \sum_i v_i^k} \tag{1}$$

where u is the network output, v is the one-hot encoding of true label, i is the voxel index, and k is the target class.

3.3   Subtyping Branch

The subtyping branch, shown as (c) in Fig. 1, shares the CNN encoding part with the segmentation decoder. We applied global average pooling on the output of convolutional layers of each resolution in the encoder, to form a 1D feature map, then used a fully connected layer with softmax function to get classification results. The loss function we used for this part is categorical cross-entropy (Eq. 2). We used 10^{-5} weight decay on all fully connected layers, as well as 0.5 rate dropout, to regularize this branch.

$$\mathcal{L}_{ce} = - \sum_{k \in K} v \, ln(u) \tag{2}$$

where u is the network output, v is the one-hot encoding of true label, and k is the target class.

To enhance this part with features extracted by radiomic methods, we took the radiomic features extracted from all four sequences of MRIs and all three labels from each patient as a 1D input vector, applied two fully connected layers with 0.01 leaky ReLU to increase non-linearity, then concatenated with features extracted from segmentation part and passed all features to the final classification layer.

3.4 Multi-task Loss Function

The segmentation branch and the classification branch were trained together, and the total loss function (Eq. 3) is a weighted sum of the two respective loss functions. After trying different weights, we found $w_1 = 1$ and $w_2 = 0.01$ worked well.

$$\mathcal{L}_{\text{total}} = w_1 \mathcal{L}_{\text{dc}} + w_2 \mathcal{L}_{\text{ce}} \tag{3}$$

4 Experiments and Results

4.1 Dataset

We use The Cancer Imaging Archive (TCIA) [8] part [2,3] of the BraTS 2018 training dataset for subtyping, in which the MRIs were manually segmented into three tumor sub-regions, namely peritumoral edema (ED), enhancing tumor (ET) and non-enhancing tumor/necrotic core (NET/NCR). For these MRIs, we managed to find the corresponding genome data from The Cancer Genome Atlas (TCGA), which was used for subtyping. Three LGG cases (TCGA-DU-A5TW, TCGA-DU-A6S7, and TCGA-DU-A6S8) from which we failed to extract radiomic features, were excluded from our experiments. Training and validation sets are randomly divided by 4 : 1 (133 vs. 31 cases). Table 1 shows the data distribution. For segmentation evaluation, we used two test sets: test set one is the non-TCIA part of the BraTS training set (118 cases), test set two is the BraTS validation set (66 cases).

Table 1. Distribution of the used dataset.

	HGG		LGG	
	IDH mutant	Wild type	IDH mutant	Wild type
Training	3	78	43	9
Validation	0	21	7	3

4.2 Training and Inference

Because the MRIs were scanned from multiple clinical centers and different devices, and images of different sequences usually have different intensity distributions, we normalized the MRI intensities by subtracting mean and dividing by standard deviation of each case, each sequence and only the brain region. For radiomic features, we normalized them by subtracting mean and dividing by standard deviation of each MRI sequence, each tumor sub-region and each feature among all cases. No data augmentation methods were applied.

We used the same training strategy for all models. In each epoch, we took two randomly cropped patches of size $128 \times 128 \times 128$ which include the tumor from each case. We used Nesterov's Accelerated Gradient (NAG) optimizer, with the momentum of 0.9 and an initial learning rate of 0.01. Learning rate was reduced by a factor of 0.1 if validation loss did not decrease in the last 30 epochs, and training was stopped if validation loss did not decrease in the last 50 epochs. We stored the weights which gives the lowest validation loss for inference.

For inference, the same normalization methods during training were applied. We also used the same patch size of $128 \times 128 \times 128$, so each MRI was divided into eight overlapping patches. The segmentation results were combined by priority $ET > NET/NCR > ED$. Then we found the largest consecutive tumor region, and used a patch with this region in the center to get a refined segmentation result and also the classification result.

All experiments were conducted on a machine with AMD Ryzen Threadripper 2950X CPU and an Nvidia RTX 2080ti GPU with 11 GB VRAM. Training took about 6 min for each epoch, and inference for each case took about 8 s.

4.3 Results

Subtyping Improved by Combining Features. All grading and subtyping classification results were evaluated on the 31-case validation set. Table 2 shows the results of grading (HGG vs LGG). We can see that both radiomics and CNN alone gave good predictions, and CNN enhanced with radiomics gave better predictions than them both. Figure 2 shows the results of subtyping (HGG vs LGG with IDH mutation vs LGG without IDH mutation). Similarly, CNN enhanced with radiomics gave better predictions than CNN alone, and better than reported radiomics-related subtyping results (0.80 for IDH-1 in Grade II Glioma by Yu 2016 [22] and 0.80 for IDH-1 in LGG by Li 2017 [13]).

Segmentation Improved by Multi-task Network. All segmentation results were evaluated via BraTS online evaluation platform, which calculates mainly Dice score and 95% Hausdorff distance over three tumor sub-regions: the enhancing tumor (ET), the tumor core (TC) which includes ET and NET/NCR, and the whole tumor (WT) which includes ET, NET/NCR and ED. Figure 3 shows the validation losses of the segmentation branch of our multi-task neural network and the segmentation-only 3D U-Net during training. It shows that our multi-task network converged to a lower loss than segmentation-only 3D U-Net.

Table 2. Classification results for HGG versus LGG.

Accuracy	Radiomics only (random forest)	CNN only	Radiomics with CNN	Cho 2017 [6]
Total	0.90	0.8387	**0.9677**	0.8981
HGG	0.91	0.8095	1	N/A
LGG	0.86	**0.9**	**0.9**	N/A

Table 3 shows the evaluation results of the segmentation from our multi-task network and segmentation-only 3D U-Net. We can see that overall our multi-task network gave better segmentation results that achieved higher Dice scores and shorter Hausdorff distances than segmentation-only 3D U-Net.

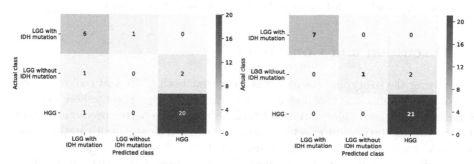

(a) CNN only, accuracy 0.8387.   (b) Radiomics with CNN, accuracy **0.9355**.

Fig. 2. Classification results for three classes.

Table 3. Segmentation results.

(a) Test set one (non-TCIA part of BraTS 2018 training set, 118 cases).

Metrics (Mean)	Base	multi-task
Dice ET	0.70513	**0.71258**
Dice WT	0.81518	**0.83550**
Dice TC	0.74622	**0.76648**
Hausdorff95 ET	**5.69995**	6.18870
Hausdorff95 WT	16.16263	**11.23476**
Hausdorff95 TC	16.36485	**14.12887**

(b) Test set two (BraTS 2018 validation set, 66 cases).

Metrics (Mean)	Base	multi-task
Dice ET	**0.73474**	0.7233
Dice WT	**0.87925**	0.87916
Dice TC	0.7464	**0.76891**
Hausdorff95 ET	**4.80279**	7.4371
Hausdorff95 WT	10.31239	**8.53958**
Hausdorff95 TC	14.00776	**12.57655**

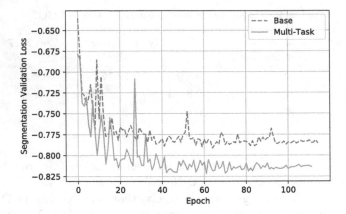

Fig. 3. Segmentation branch validation loss. Blue dash: base segmentation-only 3D U-Net, orange solid: multi-task neural network. (Color figure online)

5   Discussion and Conclusion

Comparing the subtyping results, it is clear that both CNN and radiomics captured some useful features for classification that the other method did not. Enhanced with radiomics, our multi-task network gave better predictions. Similar to the regularization effect of autoencoder path in [18], both paths in our network worked as an additional constraint for each other, improving the representation power of features extracted by the shared path, thus improved the performance of the whole network. Considering the gap in segmentation results between ours and the BraTS 2018 winner's (single model on BraTS validation set, Dice scores 0.8145, 0.9042, 0.8596 for ET, WT, TC) [18], though the small size of training dataset we used could be an important factor, there are still potential improvements in our CNN architecture.

In this work, we propose a multi-task neural network, that takes glioma MRIs and optionally radiomic features as inputs, and generates subtype predictions for the tumor as well as multi-region tumor segmentation. Enhanced with radiomics, this network achieved promising accuracy on tumor grading and subtyping based on IDH mutation: 96.77% for grading (HGG or LGG), 93.55% for subtyping (LGG with/without IDH mutation or HGG), showing the great potential of radiomics enhanced deep learning in non-invasive glioma diagnosis.

Acknowledgement. The results shown here are in whole or part based upon data generated by the TCGA Research Network: https://www.cancer.gov/tcga.

References

1. Aerts, H.J.W.L.: The potential of radiomic-based phenotyping in precision medicine. JAMA Oncol. **2**(12), 1636 (2016). https://doi.org/10.1001/jamaoncol. 2016.2631
2. Bakas, S., et al.: Segmentation labels for the pre-operative scans of the TCGA-GBM collection (2017). https://doi.org/10.7937/k9/tcia.2017.klxwjj1q
3. Bakas, S., et al.: Segmentation labels for the pre-operative scans of the TCGA-LGG collection (2017). https://doi.org/10.7937/k9/tcia.2017.gjq7r0ef
4. Bakas, S., et al.: Advancing the cancer genome atlas glioma MRI collections with expert segmentation labels and radiomic features. Sci. Data **4**, 170117 (2017). https://doi.org/10.1038/sdata.2017.117
5. Bakas, S., Reyes, M., Jakab, A., Bauer, S., Rempfler, M., Crimi, A., et al.: Identifying the best machine learning algorithms for brain tumor segmentation, progression assessment, and overall survival prediction in the BRATS challenge. arXiv e-prints. arXiv:1811.02629 (2018)
6. Cho, H.H., Park, H.: Classification of low-grade and high-grade glioma using multimodal image radiomics features. In: 2017 39th Annual International Conference of the IEEE Engineering in Medicine and Biology Society (EMBC). IEEE (July 2017). https://doi.org/10.1109/embc.2017.8037508
7. Çiçek, Ö., Abdulkadir, A., Lienkamp, S.S., Brox, T., Ronneberger, O.: 3D U-Net: learning dense volumetric segmentation from sparse annotation. In: Ourselin, S., Joskowicz, L., Sabuncu, M.R., Unal, G., Wells, W. (eds.) MICCAI 2016. LNCS, vol. 9901, pp. 424–432. Springer, Cham (2016). https://doi.org/10.1007/978-3-319-46723-8_49
8. Clark, K., et al.: The cancer imaging archive (TCIA): maintaining and operating a public information repository. J. Digit. Imaging **26**(6), 1045–1057 (2013). https://doi.org/10.1007/s10278-013-9622-7
9. Eckel-Passow, J.E., Lachance, D.H., Molinaro, A.M., Walsh, K.M., Decker, P.A., Sicotte, H., et al.: Glioma groups based on 1p/19q, IDH, and TERT promoter mutations in tumors. New Engl. J. Med. **372**(26), 2499–2508 (2015). https://doi.org/10.1056/NEJMoa1407279. pMID: 26061753
10. Gillies, R.J., Kinahan, P.E., Hricak, H.: Radiomics: images are more than pictures, they are data. Radiology **278**(2), 563–577 (2016). https://doi.org/10.1148/radiol.2015151169
11. Isensee, F., Kickingereder, P., Wick, W., Bendszus, M., Maier-Hein, K.H.: No new-net. In: Crimi, A., Bakas, S., Kuijf, H., Keyvan, F., Reyes, M., van Walsum, T. (eds.) BrainLes 2018. LNCS, vol. 11384, pp. 234–244. Springer, Cham (2019). https://doi.org/10.1007/978-3-030-11726-9_21
12. Lao, J., et al.: A deep learning-based radiomics model for prediction of survival in glioblastoma multiforme. Sci. Rep. **7**(1), 1–8 (2017). https://doi.org/10.1038/s41598-017-10649-8
13. Li, Z., Wang, Y., Yu, J., Guo, Y., Cao, W.: Deep learning based radiomics (DLR) and its usage in noninvasive IDH1 prediction for low grade glioma. Sci. Rep. **7**(1), 1–11 (2017). https://doi.org/10.1038/s41598-017-05848-2
14. Long, J., Shelhamer, E., Darrell, T.: Fully convolutional networks for semantic segmentation. In: 2015 IEEE Conference on Computer Vision and Pattern Recognition (CVPR). IEEE (June 2015). https://doi.org/10.1109/cvpr.2015.7298965

15. McKinley, R., Meier, R., Wiest, R.: Ensembles of densely-connected CNNs with label-uncertainty for brain tumor segmentation. In: Crimi, A., Bakas, S., Kuijf, H., Keyvan, F., Reyes, M., van Walsum, T. (eds.) BrainLes 2018. LNCS, vol. 11384, pp. 456–465. Springer, Cham (2019). https://doi.org/10.1007/978-3-030-11726-9_40

16. Menze, B.H., Jakab, A., Bauer, S., Kalpathy-Cramer, J., Farahani, K., Kirby, J., et al.: The multimodal brain tumor image segmentation benchmark (BRATS). IEEE Trans. Med. Imaging **34**(10), 1993–2024 (2015). https://doi.org/10.1109/tmi.2014.2377694

17. Milletari, F., Navab, N., Ahmadi, S.A.: V-net: fully convolutional neural networks for volumetric medical image segmentation. In: 2016 Fourth International Conference on 3D Vision (3DV). IEEE (October 2016). https://doi.org/10.1109/3dv.2016.79

18. Myronenko, A.: 3D MRI brain tumor segmentation using autoencoder regularization. In: Crimi, A., Bakas, S., Kuijf, H., Keyvan, F., Reyes, M., van Walsum, T. (eds.) BrainLes 2018. LNCS, vol. 11384, pp. 311–320. Springer, Cham (2019). https://doi.org/10.1007/978-3-030-11726-9_28

19. Ronneberger, O., Fischer, P., Brox, T.: U-Net: convolutional networks for biomedical image segmentation. In: Navab, N., Hornegger, J., Wells, W.M., Frangi, A.F. (eds.) MICCAI 2015. LNCS, vol. 9351, pp. 234–241. Springer, Cham (2015). https://doi.org/10.1007/978-3-319-24574-4_28

20. Tian, Q., Yan, L.F., Zhang, X., Zhang, X., Hu, Y.C., Han, Y., et al.: Radiomics strategy for glioma grading using texture features from multiparametric MRI. J. Magn. Reson. Imaging **48**(6), 1518–1528 (2018). https://doi.org/10.1002/jmri.26010

21. Wang, X., et al.: Machine learning models for multiparametric glioma grading with quantitative result interpretations. Front. Neurosci. **12**, 1046 (2019). https://doi.org/10.3389/fnins.2018.01046

22. Yu, J., Shi, Z., Lian, Y., Li, Z., Liu, T., Gao, Y., et al.: Noninvasive IDH1 mutation estimation based on a quantitative radiomics approach for grade II glioma. Eur. Radiol. **27**(8), 3509–3522 (2016). https://doi.org/10.1007/s00330-016-4653-3

23. Zhang, B., Chang, K., Ramkissoon, S., Tanguturi, S., Bi, W.L., Reardon, D.A., et al.: Multimodal MRI features predict isocitrate dehydrogenase genotype in high-grade gliomas. Neuro-Oncology **19**(1), 109–117 (2016). https://doi.org/10.1093/neuonc/now121

24. Zhang, X., Yan, L.F., Hu, Y.C., Li, G., Yang, Y., Han, Y., et al.: Optimizing a machine learning based glioma grading system using multi-parametric MRI histogram and texture features. Oncotarget **8**(29), 47816 (2017). https://doi.org/10.18632/oncotarget.18001

25. Zhou, C., Chen, S., Ding, C., Tao, D.: Learning contextual and attentive information for brain tumor segmentation. In: Crimi, A., Bakas, S., Kuijf, H., Keyvan, F., Reyes, M., van Walsum, T. (eds.) BrainLes 2018. LNCS, vol. 11384, pp. 497–507. Springer, Cham (2019). https://doi.org/10.1007/978-3-030-11726-9_44

Author Index

Printed in the United States
By Bookmasters